Fundamentals of Instrumentation

Fundamentals of Instrumentation

NJATC

THOMSON

™

DELMAR LEARNING

Australia   Canada   Mexico   Singapore   Spain   United Kingdom   United States

Fundamentals of Instrumentation
NJATC

Vice President, Technology and Trades SBU:
Alar Elken

Executive Director, Professional Business Unit:
Gregory L. Clayton

Product Development Manager:
Patrick Kane

Executive Marketing Manager:
Beth Lutz

Channel Manager:
Erin Coffin

Marketing Coordinator:
Penelope Crosby

Production Director:
Mary Ellen Black

Production Manager:
Larry Main

Senior Project Editor:
Christopher Chien

Art/Design Coordinator:
Francis Hogan

Editoral Assistant:
Sarah Boone

COPYRIGHT © 2005 by National Joint Apprenticeship Training Committee

Printed in the United States of America
1 2 3 4 5 XX 07 06 05 04

For more information contact
Delmar Learning
Executive Woods
5 Maxwell Drive, PO Box 8007,
Clifton Park, NY 12065-8007
Or find us on the World Wide Web at
www.delmarlearning.com

All rights reserved. No part of this book may be reproduced in any form or by any means, stored in a retrieval system, transmitted by any means, electronic, mechanical, photocopying, recording, or otherwise, without permission in writing from the National Joint Apprenticeship Training Committee (NJATC). No patent liability is assumed with respect to the use of the information contained within this book. While every precaution has been taken in preparation of this book, the NJATC and the author assume no responsibility for errors or omissions. Neither is any liability assumed by the NJATC, the author, or any of the manufacturers of the test equipment used for explanatory purposes for damages resulting from the use of the information contained herein. Every effort has been made to explain the safe usage of electrical test equipment, but it is impossible to predict in every instance how and for what purpose equipment may be utilized. The operator of any equipment should always refer to the instruction manual of the test set to be used and observe all the safety precautions therein listed.

For permission to use material from the text or product, contact us by
Tel.   (800) 730-2214
Fax   (800) 730-2215
www.thomsonrights.com

Library of Congress Cataloging-in-Publication Data:

Stafford, Todd.
    Fundamentals of instrumentation / Todd Stafford.--1st ed.
        p. cm.
    ISBN 1-4018-9789-4
    1. Mensuration. 2. Measuring instruments. 3. Electric meters. I. Title
    T50.S73 2004
    629.8--dc22
                            2004022060

NOTICE TO THE READER

Publisher does not warrant or guarantee any of the products described herein or perform any independent analysis in connection with any of the product information contained herein. Publisher does not assume, and expressly disclaims, any obligation to obtain and include information other than that provided to it by the manufacturer.

The reader is expressly warned to consider and adopt all safety precautions that might be indicated by the activities herein and to avoid all potential hazards. By following the instructions contained herein, the reader willingly assumes all risks in connection with such instructions.

The publisher makes no representation or warranties of any kind, including but not limited to, the warranties of fitness for particular purpose or merchantability, nor are any such representations implied with respect to the material set forth herein, and the publisher takes no responsibility with respect to such material. The publisher shall not be liable for any special, consequential, or exemplary damages resulting, in whole or part, from the readers' use of, or reliance upon, this material.

Contents

CHAPTER 1
Fundamentals of Control Systems 2

CHAPTER 2
Instrument Symbols and Identifiers 12

CHAPTER 3
Fundamentals of Calibration 18

CHAPTER 4
Fundamentals of Pressure   32

CHAPTER 5
Fundamentals of Flow   46

CHAPTER 6
Fundamentals of Liquid Level   70

CHAPTER 7
Fundamentals of Temperature   90

CHAPTER 8

Fundamentals of Pneumatics and Control Valve Actuators 110

CHAPTER 9

Fundamentals of Controllers 126

CHAPTER 10
Fundamentals of Control   136

CHAPTER 11
Fundamentals of Analytical pH Measurement   152

CHAPTER 12
Fundamentals of Smart Instrument Communicators   160

CHAPTER 13
Fundamentals of Smart Instrument Calibration   170

CHAPTER 14
Fundamentals of Instrument Installation   182

CHAPTER **15**

Fundamentals of Instrument Maintenance 196

CHAPTER **16**

Fundamentals of Control Valve Maintenance 208

CHAPTER 17
Fundamentals of Instrument Tubing 232

CHAPTER 18
Fundamentals of Documentation 240

CHAPTER 19
Fundamentals of Safety in the Process Environment **258**

Appendix A—Instrumentation and Controls Symbology **267**

Appendix B—NJATC Instrumentation and Process Control Training System **281**

Glossary **282**
Index **293**

Preface

The National Joint Apprenticeship and Training Committee (NJATC) is the training arm of the International Brotherhood of Electrical Workers (IBEW) and the National Electrical Contractors Association (NECA). Established in 1941, the NJATC has developed uniform standards that are used nationwide to train thousands of qualified men and women for demanding and rewarding careers in the electrical and telecommunications industry. To enhance the effectiveness of this mission, the NJATC has partnered with Thomson Delmar Learning to deliver the very finest in training materials for the electrical profession.

Fundamentals of Instrumentation is a resource for understanding the concepts and implementation of instrumentation. The text applies electrical principles that you have learned in an electrical apprenticeship or other training program. Ohm's Law and other electrical principles are not discussed, but they are applied where they affect field devices and installation requirements. The text introduces relevant instrumentation principles required for accurate measurement and control and assumes a prior understanding of the basic electrical principles and installation requirements.

Instrumentation is broadly defined as any device that performs a measuring or controlling function. Rather than trying to define, describe, and study the numerous devices that are used for measuring or controlling process systems, the text identifies and defines the physical properties that affect all devices, allowing for the installation, calibration, and use of all devices. The installation requirements for process measurement devices require consideration and understanding of the process physics being measured. The text covers the physical properties you must consider when installing a device.

Also, each device has a "calibration procedure" that you must follow to accurately calibrate. The calibration procedure is often "designed" by the manufacturer. The text provides in-depth knowledge of the parameters that you must adjust or adapt to achieve accuracy in a field measurement device regardless of the make and model of the device. The calibration errors that you may find within a device are clearly discussed and you will learn that different manufacturers have specific requirements for eliminating inaccurate signals.

There are four basic steps that must be covered to adequately study the field of instrumentation. They are mounting (installation), wiring, process connections (impulse tubing), and calibration. These steps are the backbone of knowledge for those who consider themselves proficient in "instrumentation." Approximately 85 % of instrumentation work-hours are spent on these four narrowly defined categories. Therefore, the text is limited to explaining and preparing you to execute the requirements to satisfy these four groups.

Instruments that record process variables and devices or provide methods to control processes were utilized in process "systems" before the industrial revolution of the 1800s. These devices were *pneumatic* and used the physical laws of pressure to define and measure process variables. The discovery of the transistor and integrated circuit technologies in the mid-1900s provided a method to create and implement instruments to perform the same functions as their earlier pneumatic counterparts, but with increased accuracy and reliance at a reduced cost. By the late 1900s, instrumentation had become primarily *electronic*. Although nearly all devices installed today are electronic, there are a few industries and applications in which pneumatic installations are required or are more efficient. The text concentrates on using electronic devices for instrumentation, but does provide information on pneumatic control devices.

Regardless of the device you use to measure a process, electronic or pneumatic, the process interface, called the primary element in many instances, has not changed. You can study the physical properties without defining the measurement device that records the process. To correctly and accurately install and calibrate the device, you must understand how the impulse tubing carries the process being measured to the device performing the measurement function. Likewise, the final control elements in an automatic control system still use pneumatics to position control valves using pressure forces.

Chapter 1 describes process systems and explains how devices are used in order to provide a method of automatic control. By examining a particular *loop* or function of the process control system, you learn that you are responsible for the accurate control of a device you are learning to calibrate and install. Seeing how the devices are used is critical. You must always keep a *goal* in sight and reach for that goal. The goal of working devices in an automatic control system is to help you understand how the devices perform in a working process.

Chapter 2 provides basic information on the symbols and identification numbers used to identify specific instruments. The chapter helps you begin to understand the physical layer consisting of wiring methods and loop *tags* that are specific for instruments. Drawings and notations that you will use are covered. The symbols, notes, and identification methods used to present documented calibrations, mounting requirements, and control parameters use these same identifiers when related to actual instruments.

Chapter 3 begins with the calibration process. In theory, the calibration process can be performed without knowing the conditions under which the device will perform when installed. The concepts of zero and span are discussed as well as the other instrument errors you may see in instrumentation. The errors are presented without relating them to a particular process for it is advantageous for you to be able to identify errors in an instrument under any process, measurement setup. The remaining chapters will further refine how the instrument errors that you may discover in a device have been created due to a process physics relation or installation.

Chapter 4 begins to look at the physical properties that are measured. Pressure is the most fundamental of all process conditions, and indeed pressure can be used to infer a measurement property or can be measured directly. You will see that pressure can be used to measure flow, level, and temperature, so it is critical that you learn the fundamentals presented here.

Chapter 5 introduces the fundamentals of flow and how physical properties of a moving fluid can be measured and interpreted to reflect a flow rate. Orifice plate measurements, volumetric flow rates, and Coriolis principles are discussed along with many others. Bernoulli's principle is presented so you can begin to understand the effects of a moving fluid upon the calibrated device.

Chapter 6 presents the fundamentals of liquid level. This chapter uses information presented in earlier chapters, such as pressure, to derive levels. Ultrasonic measurements, radar, and Archimedes Law are all stated and explain how devices use these principles to obtain level measurements.

Chapter 7 provides the last of the four physical parameters. The fundamentals of temperature enable you to understand the different temperature scales you may be confronted with. The text also provides methods to convert between the different scales. This chapter is unique in that temperature is the only physical property that cannot be measured directly. Instruments can only measure the effects of temperature upon objects and the text explains how the measurement process differs.

Chapter 8 provides a brief introduction to the use of pneumatics and the physical properties that are used to measure and control a process. The predominant force used in controls today is pneumatic pressure. The use of pneumatics for measuring specific pressures and the use of pneumatic pressures to position a control valve are thoroughly discussed.

All automatic control systems must have some form of controller that maintains order in the automatic control system. Chapter 9 provides a look at the controller that is performing the brains of the operation as well as the requirements you must provide to a controller for efficient operation.

Chapter 10 takes the information provided in earlier chapters and explains the control process. PI&Ds control is discussed to give you insight into the control process. This chapter is intended to introduce you to the process used by a controller to control the process.

Some instruments do not use the physical properties outlined previously, but use chemical properties to derive a process measurement. Chapter 11 provides information about one of the most common types of analyzers that you will use and install. pH measurements require a different setup and calibration procedure, and this process is discussed.

The increasing complexity of communications between smart devices and communications devices requires an in-depth study into the use of hand-held communications equipment. Chapter 12 references one specific type of communicator, Rosemount 275, as an example, but you can apply the concepts to all communication equipment you use to configure and calibrate smart devices.

Chapter 13 presents additional methods of calibrating smart devices. Although the process measurement properties have not changed when using smart devices to measure a process, the calibration procedures have changed. The introduction of smart devices in the field requires a different calibration procedure as well as some parameters that are introduced here.

The next three chapters, Chapters 14, 15, and 16, provide the physical relationships that you must take into account when installing or working on a device to measure or control a process. Maintenance concerns, physical mounting requirements, troubleshooting tips, and more are provided in order to ensure that devices calibrated correctly are not affected in any way.

Chapter 17 provides a look at the remaining element in the field of instrumentation. The process tubing that carries the process to the measuring device is discussed in this chapter and provides a detailed explanation of its requirements. Because you could still introduce errors into a device through incorrect installation of process tubing, it is critical that you cover this information completely.

Chapter 18 provides a detailed look into the documentation packages that are used to build and maintain an automatic control system. The hierarchy of the documents is given as well as how to determine the applicable drawing that is to be used for a reference.

Chapter 19 provides OSHA Safety and Health Construction excerpts that all workers in the field of instrumentation need to be familiar with. This is not a comprehensive list but rather a list of the most commonly used precautions. This chapter tries to present safety as a work requirement rather than a hindrance.

Appendix B consists of 9 loop sheet figures that are located on the CD enclosed with this text. For your convenience in training, both DWG and PDF versions of these files are included to best suit your needs.

NJATC ACKNOWLEDGMENTS

NJATC Technical Writer and Editor, Todd W. Stafford, NJATC Senior Director

ADDITIONAL ACKNOWLEDGMENTS

The NJATC would additionally like to thank the following individuals for their help and support: James Dawig and Gerald Boyer, International Brotherhood of Electrical Workers, and Brad King, United Association of Plumbers, Pipefitters and Steamfitters.

chapter 1

Fundamentals of Control Systems

■ OUTLINE

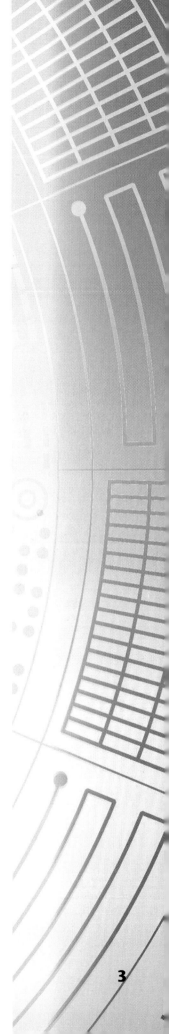

■ OVERVIEW

All **automatic control systems** can be broken down into specific devices and loops. The function of a control system can be similarly broken down as well. By identifying the basic control devices and loops, the operation of the system as a whole can be understood. This chapter provides needed information to understand and comprehend the devices utilized in a control system, as well as providing a brief description of the basic configuration of a control system.

■ OBJECTIVES

After completing this chapter, you should be able to:

- Describe the purpose of a control system.
- Describe the process variables that can be measured in a control system.
- Define automatic control.
- Describe an analog signal.
- Describe a discrete signal.
- Understand signal paths for control signals.
- Identify operator's control interface(s).
- Provide detailed information as to the importance of field instrumentation.

■ INTRODUCTION

1.1 Defining Control Systems

Control systems are used to regulate a process, monitor a process, or indicate when a process has reached a desired result (a **setpoint**). In process control, the basic goal is to automatically regulate a process at a predetermined value. This is the primary purpose of a **control loop**. A control loop begins by measuring a process. A signal is then sent to a **controller** that determines some form of control through the use of **software.** Lastly, a signal is sent to a final control element that executes a control action.

1.2 Defining Process Variables

A **process** is defined as any function or operation utilized in the treatment of a material. For example, the operation of adding heat to water is a process. Processes are taking place wherever you go. Most of us work with the basics of process and process control every day. **Process variables** include flow rate, level, temperature, etc. As you drive a car, you are performing process control functions by controlling the speed of the car as well as its direction through steering. As a further example of our interaction with process control, think about the building you are in. Temperature and ventilation control provide an adequate environment for our comfort with minimal interaction from us. Process control is the method by which we regulate a particular process. We perform process control when we vary the gas flow into our automobile engine and adjust a thermostat for environmental comfort. We will further develop our understanding of process control by first examining the components that enable us to control a working process (Figure 1–1).

2-17

FIGURE 1–1 Industrial process facilities often contain several thousand field input/output, all of which must be installed and calibrated correctly.

1.3 Historical Background

In the not so distant past, if you wanted to fill a tank to a certain level, you would watch the rising liquid. When you achieved the desired level, you would manually halt the filling process. This was the first form of process control.

For this example, the object of automatic process control is to perform the filling process automatically without any help from a human operator. To provide automatic control, the process system is modified or adapted to allow machines and devices to perform the controlling process. The control system would consist of several instruments that would initiate the record and regulate the filling process. Therefore, the controller would take the place of the human decisions that are usually made to ensure an efficient process. The instruments could be measuring the level of the tank's liquid, the flow of the liquid into the tank, the pressure or weight of the liquid stored in the tank, and the tank's liquid temperature.

1.4 Automatic Control

Depending upon the control variables that are defined, some or all of the process variables (flow rate, level, temperature, etc.) may be needed for the controller to make decisions that would be required to accurately fill and maintain the proper tank level. The decisions normally "made" by an **automatic controller** regulating a process can be performed by a variety of methods, but the most common method today is to allow control to be executed by a microprocessor-based controller (Figure 1–2).

2-10

Microprocessor-based control systems are capable of executing a wide variety of user-defined **algorithms,** or instructions, that initiate

FIGURE 1–2 A control room is the central location for viewing and controlling the process facility.

and control the various processes. The **control system,** defined as controlling means governed by the microprocessor, is the collection of the components needed to perform the function of maintaining an efficient process. Depending upon the system in question, the user may define various inputs and outputs (I/O) that will be monitored or controlled by the microprocessor. The I/O defined are the field instruments that gather process data (**inputs**) and transmit the data to the controller in the form of a signal, as well as the signals sent to control the process (**outputs**) that are used to position a final control element.

■ CONTROL SIGNALS

1.5 Analog Signals

Typical signals used in process control are analog, discrete, and digital. An **analog** signal usually represents a range of values read by a sensor. This analog signal is usually represented by 4–20 mA. Analog signals can be used as inputs, recording a range of values, and can also be used in the output signal by varying a position of a final control element (Figure 1–3). When devices are calibrated correctly, the process measurement (input) is accurately represented by the 4–20 mA signal and the output signal correctly positions an element for control.

1.6 Discrete Signals

Another form of signal used is the **discrete** signal. The discrete signal is much easier to implement, but provides less information than the analog signal. For example, with a level switch, the switch contact closes when a certain level in a tank is measured. This is referred to as "making" and allows the signal to flow. The "making" of the switch contacts is the only indication of what level is in the tank. A discrete signal can also be used to control a process; for example, it can be used

FIGURE 1–3 A marshalling panel connects field wiring to controller termination points. I/O are typically assigned a rack, slot, box, or point assignment to terminate field wiring.

2·20

to turn on or off a pump or other on/off devices to initiate control. An example of discrete control is the application of a sump pump with high and low switches that turn a pump on or off depending upon the level in the sump.

1.7 Signal Transmission

Once the I/O are defined and configured in the controller, they are capable of transferring information via a signal from the field devices to the controller, and the appropriate response is taken or observed. The transmission of the signals to and from a controller is most often accomplished through the use of twisted-pair wiring. *Twisted-pair wiring* is a physical construction of two current-carrying conductors and bare ground. The wires are twisted around each other to negate capacitive and inductive effects upon the current flowing in the conductors. The twisted pair is surrounded by a shield to further insulate and isolate

FIGURE 1–4 Boiler house fire eye receivers.

the conductors. The twisted-pair wiring path creates a signal loop that allows the transfer of information from measuring devices to a controller to the final control element (Figure 1–4). Once a controller receives I/O, the transfer of information between the controller and the display used by operations to control the process may be executed internally or transmitted via a dedicated highway **interface.**

Communication lines are often redundant (more than one) to allow for the repair and maintenance of one line and to also provide a path for checking the information as it is received by the controller. Depending upon the level of control, software is used for diagnostic capabilities to ensure the process is performing as smoothly as the controller is directing.

The most important bit of information gathered by the controller is the information received from field instruments that determine the appropriate response for the working process. This information is gathered by field-measuring devices that should accurately portray the working process. It is important for field technicians to understand the working concepts of a system to know how their devices can interface with and change the operating algorithm of a controlled process. As you have learned previously, control systems can contain a variety of devices and/or controllers that are used to control processes.

The common thread ties all automatically controlled working processes of initiating and final control elements of a system together. An **operating system** depends upon the accurate sensing, signaling, and transmission of data—these are the duties of the field devices and, therefore, are the responsibility of the field technician. In order for field technicians to properly calibrate, troubleshoot, and loop check various systems, they must have a working knowledge of the control systems that are used to control the various processes. Each field device that is used to transmit a **variable** to the controller must be accurate; hence the importance of understanding the steps of calibration and installation of measuring devices (Figure 1–5). Field devices provide process measurements to the controller and final control elements receive a signal from the controller for position to control the process.

FIGURE 1–5 Gas control station for thermal oxidizer.

■ OPERATOR INTERFACE

1.8 Graphic User Interface

Once field inputs have transferred information to the controller, operations control **monitors** provide a visual method to "see" the process. Video displays, the most common and visible components in a system, are often referred to by a number of names including Graphic User Interface (GUI), Human Machine Interface (HMI), and Man Machine Interface (MMI). In industrial applications, operators use displays to gather information that allows them to see the process. A process can include many variables that may have the option of being monitored or used for control. With automatic control systems, the I/O can be viewed at various locations on displays via **communications** highways that allow a signal operator to view multiple processes. Often, several locations will have a display for monitoring I/O, while only one or two locations may be available for control functions. A display allows for easy viewing of the process through information gathered from field devices. An entire process facility can often be viewed from a single machine with multiple views (Figure 1–6).

1.9 Control Devices

The components that are monitored by a video screen are actually devices that are used to measure a particular function, or devices that respond to a particular control command given by the controller. If a video display recorded the RPM of rotating machinery in a controlled process, do you think the signal received by the controller would be sent by the rotating machinery or by a measuring/recording device monitoring the rotating machinery? Of course, the device that is used to count the number of revolutions of the rotating machinery is the device sending the signal to the video display. In addition, if the rotating speed of the machinery was needed to increase its RPM, do you think

FIGURE 1-6 Boiler house control room screen display. Displays such as this one provide a method for operating personnel to view and manipulate the process.

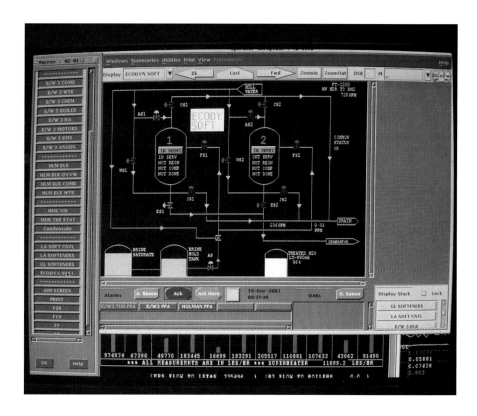

a signal would be sent to the machine or to a device that controls the rotation speed? The device that controls the rotation speed is the final control element. These are simplified examples of field devices (or instruments) that are used to control a process.

For all processes, the signals used to record, monitor, and control are performed by devices that provide an interface with a controller to a working process. This is another reason why field technicians must understand how their devices reflect a working process.

1.10 Shared Control

Out of necessity, a controller and its associated display sometimes perform the control functions for several processes. This controller, referred to as a shared controller, contains a user-defined algorithm that is changeable to permit flexibility for the user. The controller's algorithm is roughly a listing of instructions to be executed by the controller and often contains diagnostic features to verify information gathered by the system.

A process must be monitored continuously for accurate automatic control. The algorithm contained in the controller determines the frequency at which the various processes of I/O are scanned. The scan of a program, or algorithm, is the repetitive reading of the I/O field to execute instructions contained in the program. Flow rates, levels, temperatures, and pressures, among others, are all read by a controller and the total variables for each system may number in the several hundreds. The outputs of a controller to valves, motor drives, solenoids, heating elements, etc. are equally adjusted or maintained through each scan of the controller.

The reading of the process variables, output signal adjustments, diagnostic variables, and system informational statistics can be executed as quickly as a few milliseconds and then the resulting control actions are taken. One reading and the resulting execution of the instructions entered into the controller are referred to as a scan. Now that you have learned that the scan time of a program can be very fast and that control actions are taken from the readings obtained from the scan, it is easy to see the importance of having an accurate sensing and signaling device sending information to a controller (Figure 1–7).

■ KEY ELEMENTS

1.11 Understanding the Purpose of Instrumentation

Field technicians must realize that they may be asked to troubleshoot, calibrate, terminate, and perform any of a number of adjustments to field I/O. Calibration is only a part of working with instruments. Someone who wants to work on instrumentation must be capable of understanding and performing a variety of tasks including calibration, mounting, terminating, and processing (impulse) tubing principles.

The one common denominator to all of the devices in a control system is that they are shown on drawings that can display the location, calibration, type, etc. All of the devices that are used by controllers are diagrammed on drawings that show how these components interact with each other. It is important for you, the journeyperson, to understand instruments, control symbols, and identification standards so

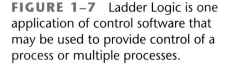

FIGURE 1–7 Ladder Logic is one application of control software that may be used to provide control of a process or multiple processes.

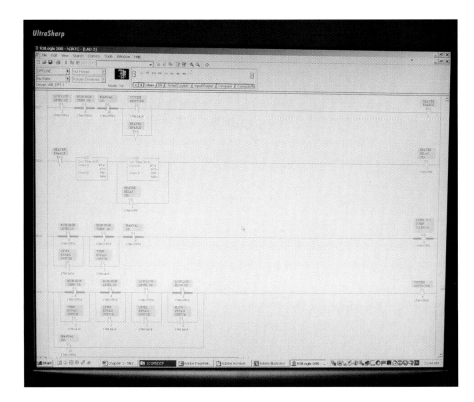

that you can interpret how the devices are connected to the system and what data are to be sent to the controller. When field technicians understand how a control system functions and how a sensing device gathers information and signals a controller, they may consider themselves proficient in the fundamentals of how instruments sense and transmit process variables. This is the beginning of understanding the fundamentals of instrumentation.

■ SUMMARY

Instrumentation is the method by which a process variable is obtained in a usable form to position a final control element to control a process with minimum deviation. This chapter has outlined a basic control loop that portrays the functionality of each device in a control loop. The types of signals were discussed as well as the use of each signal. More detailed information is forthcoming to further define the control signals utilized. Use the information gathered here to build upon. Each subsequent chapter will provide additional information on the building blocks defined in this chapter.

■ REVIEW QUESTIONS

1. What is the purpose of an automatic control system? What are the devices used in a typical control loop that are needed to have an automatic control system operate correctly?

2. What are the four basic process variables that can be measured by field instruments?

3. What are the two basic control signals defined in Chapter 1 and give a brief explanation of each?

4. What is the purpose of an operator's control station?

chapter 2

Instrument Symbols and Identifiers

■ OUTLINE

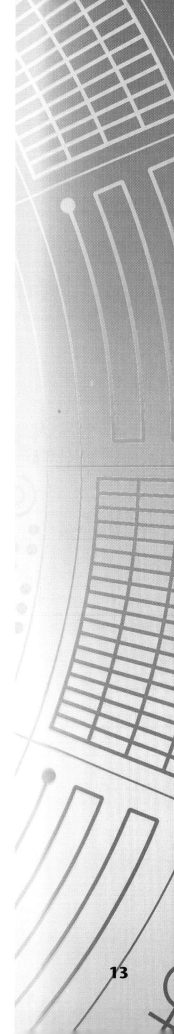

■ OVERVIEW

Instruments that are used in process control systems are identified with a unique identification code, usually called a **Tag Number.** The unique code allows for a function of a device to be determined by interpreting the Tag Number on a drawing or specification sheet. The Tag Number can provide information on how the device operates, its control purpose, and the loop designation it receives. Also, a device's Tag Number can include the process connections, control or monitoring purpose, and signal path. Commonly, instrument "tags" are shown on a **P&ID** (Process and Instrumentation Diagram) as well as "loop sheets" used by field personnel.

■ OBJECTIVES

After completing this chapter, you should be able to:

- Identify the line symbols shown on instrument drawings.
- Identify the instrument and function symbols used on instrumentation drawings.
- Use a device Tag Number to identify a device's function and loop number.
- Describe the typical drawings used for instrumentation.
- Interpret typical letter combinations for Tag Numbers.

■ INTRODUCTION

2.1 Line, Instrument, and Function Symbols

Symbols portrayed on diagrams and drawings are used to depict instrumentation used for control of process systems (Figure 2–1). Symbols can depict or represent many applications of a control system, such as discrete instruments, analog instruments, primary elements, final control elements, termination points, and signal transmission type. Function blocks are also used on the same diagrams or drawings to represent the functional design of the control loop (Figure 2–2). Function blocks describe the operational intent of the control loop rather than individual instruments.

Symbol size may vary according to the user's demands. Inaccessible or behind-the-panel devices or functions are usually shown with a dashed line.

2.2 Instrument Identification

Each instrument and/or function has an identifier label attached to it when shown on associated instrument drawings. The identifier, called a tag, contains an alphanumeric string and is determined by standards that are used for instrument identification. The loop connection number is common to all instruments and connections to the loop. The loop number often has a prefix and/or a suffix, or both, to complete the task of identifying an associated device. The following text shows a typical Tag Number and the meaning of each letter and numeral:

FIC   013—Instrument identification or Tag Number

F   013—Loop identification

       013—Loop number

FIGURE 2–1 Instrument line symbols and identifiers.

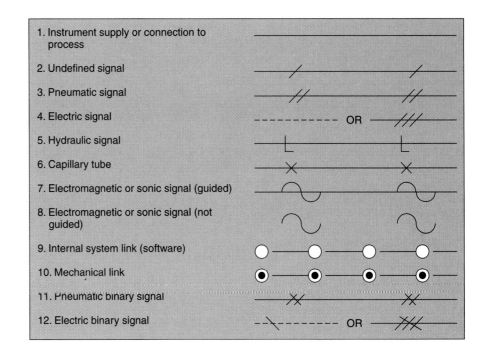

1. Instrument supply or connection to process	
2. Undefined signal	
3. Pneumatic signal	
4. Electric signal	OR
5. Hydraulic signal	
6. Capillary tube	
7. Electromagnetic or sonic signal (guided)	
8. Electromagnetic or sonic signal (not guided)	
9. Internal system link (software)	
10. Mechanical link	
11. Pneumatic binary signal	
12. Electric binary signal	OR

FIGURE 2–2 Instrument and function symbols.

	Primary location normally accessible to operator	Field mounted	Auxiliary location normally accessible to operator
Discrete instruments			
Shared display Shared control			
Computer function			
Programmable logic control			

FIC —Functional identification

F —First letter

IC —Succeeding letters

As explained previously, some controllers may control the process for more than one unit. The instrument tag is used as an identifier for the controlled elements. If the controller is performing the control functions for more than one area, a Tag Number may also be present in another area. By using the expanded Tag Number, a Tag Number can be present more than once. An expanded tag will have a hyphen that is used to designate a certain area. The following text shows a typical Tag Number and the specific meaning of each letter and numeral:

11-FAH-013X —Expanded Tag Number

11 —Optional prefix (used to designate area or unit)

X —Optional suffix (to further identify a device)

The use of hyphens in the expanded Tag Number is optional.

2.3 Instrumentation Documents (Documentation)

Proper interpretation of documentation is as critical as the actual installation of devices. The following is a description of possible documentation you may encounter:

1. P&ID — Process and Instrumentation Diagram; sometimes called Piping and Instrumentation Diagram. Shows process-related piping and the instrumentation. Supersedes all other documentation.

2. Loop sheet (Figure 2–3) — Shows all related (loop number) devices for controlling means and the wiring path, termination points, and device locations.

FIGURE 2–3 An example of a loop sheet. Notice the device tag, terminations, grounding requirements, input loop, output loop, and local indicator.

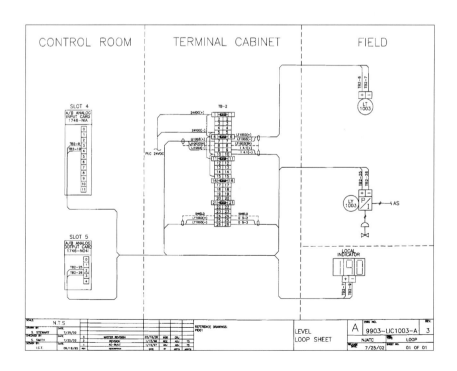

3. Specification sheet (spec sheet) — Record of devices listed by Tag Number showing related ranges, setpoints, material construction, process exposure, and related environmental variables.

4. Wiring diagram — Shows related devices between loop sheets.

5. Elementary drawing — Indicates termination points of controlling means.

2.4 Typical Letter Combinations

Typical letter combinations for field devices allow for accurate identification of a single device by representing the device by a known reference standard. Utilizing a readily available sequence of letters to identify a device, a standard allows for accurate portrayal of the device on an instrument diagram or drawing (Figure 2–4). Typical letter combinations provide the easiest method to transfer information from the drawings used to build or maintain a control system to those who are performing the actual work upon the system. There are instances in which systems are constructed and maintained without a typical letter combination. Such occasions demand more reliance on the field worker to interpret and identify devices and functions of the device.

■ SUMMARY

Identification of field devices is critical to installation and maintenance requirements associated with field I/O. Being able to use the Tag Numbers associated with instruments allows for proper identification of devices and for accurate interpretation as to the function of the device. Proper identification also minimizes mistakes in the field when multiple I/O are present.

■ REVIEW QUESTIONS

1. A unique identification code for devices shown on instrumentation drawings and diagrams is referred to as what?

2. Symbols portrayed on diagrams and drawings are used to depict what?

3. Draw in the space below an instrument line symbol that represents an electrical or electronic signal.

4. What does the symbol below represent?

5. What type of drawing shows all related (loop number) devices for controlling means and the wiring path, termination points, and device locations?

6. A device that is identified as TT1004 can be interpreted to be what?

7. A device that is identified as PSL 2345 can be interpreted to be what?

8. What does a Tag Number LAL1003 represent?

1st letter	measured variable	Controllers recording	indicating	blind	self actuated valves	readout devices recording	indicating	switches and alarm devices high	low	comb	transmitters recording	indicating	blind	solenoids, relays, computing devices	primary element	test point	well or probe	viewing device, glass	safety device	final element
A	analysis	ARC	AIC	AC		AR	AI	ASH	ASL	ASHL	ART	AIT	AT	AY	AE	AP	AW			AV
B	burner combustion	BRC	BIC	BC		BR	BI	BSH	BSL	BSHL	BRT	BIT	BT	BY	BE		BW	BG		BZ
C	users choice																			
D	users choice																			
E	voltage	ERC	EIC	EC		ER	EI	ESH	ESL	ESHL	ERT	EIT	ET	EY	EE					EZ
F	flow	FRC	FIC	FC	FCV, FICV	FR	FI	FSH	FSL	FSHL	FRT	FIT	FT	FY	FE	FP		FG		FV
G	users choice																			
H	hand		HIC	HC				HS												HV
I	current	IRC	IIC	IC		IR	II	ISH	ISL	ISHL	IRT	IIT	IT	IY	IE					IZ
J	power	JRC	JIC	JC		JR	JI	JSH	JSL	JSHL	JRT	JIT	JT	JY	JE					JV
K	time	KRC	KIC	KC	KCV	KR	KI	KSH	KSL	KSHL	KRT	KIT	KT	KY	KE					KV
L	level	LRC	LIC	LC	LCV	LR	LI	LSH	LSL	LSHL	LRT	LIT	LT	LY	LE		LW	LG		LV
M	users choice																			
N	users choice																			
O	users choice																			
P	pressure	PRC	PIC	PC	PCV	PR	PI	PSH	PSL	PSHL	PRT	PIT	PT	PY	PE	PP			PSV, PSE	PV
PD	pressure differential	PDRC	PDIC	PDC	PDCV	PDR	PDI	PDSH	PDSL		PDRT	PDIT	PDT	PDY	PE	PP				PDV
Q	quantity	QRC	QIC			QR	QI	QSH	QSL	QSHL	QRT	QIT	QT	QY	QE					QZ
R	radiation	RRC	RIC	RC		RR	RI	RSH	RSL	RSHL	RRT	RIT	RT	RY	RE		RW			RZ
S	speed frequency	SRC	SIC	SC	SCV	SR	SI	SSH	SSL	SSHL	SRT	SIT	ST	SY	SE					SV
T	temperature	TRC	TIC	TC	TCV	TR	TI	TSH	TSL	TSHL	TRT	TIT	TT	TY	TE	TP	TW		TSE	TV
TD	temperature differential	TDRC	TDIC	TDC	TDCV	TDR	TDI	TDSH	TDSL		TDRT	TDIT	TDT	TDY	TE	TP	TW			TDV
U	multi variable					UR	UI							UY						UV
V	vibration machinery analysis					VR	VI	VSH	VSL	VSHL	VRT	VIT	VT	VY	VE					VZ
W	weight	WRC	WIC	WC	WCV	WR	WI	WSH	WSL	WSHL	WRT	WIT	WT	WY	WE					WZ
WD	weight differential	WDRC	WDIC	WDC	WDCV	WDR	WDI	WDSH	WDSL		WDRT	WDIT	WDT	WDY	WE					WDZ
X	unclassified																			
Y	event state presence		YIC	YC		YR	YI	YSH	YSL			YT		YY	YE					YZ
Z	position dimension	ZRC	ZIC	ZC	ZCV	ZR	ZI	ZSH	ZSL	ZSHL	ZRT	ZIT	ZT	ZY	ZE					ZV
ZD	guaging deviation	ZDRC	ZDIC	ZDC	ZDCV	ZDR	ZDI	ZDSH	ZDSL		ZDRT	ZDIT	ZDT	ZDY	ZDE					ZDV

FIGURE 2–4 Typical letter combinations.

chapter 3

Fundamentals of Calibration

◼ OUTLINE

■ OVERVIEW

Calibration is the key requirement in ensuring proper control accuracy. The process of calibrating an instrument is also the primary objective for all work performed in the field of instrumentation. It is essential that the calibration procedure identifies calibration errors, properly records the errors identified, requires that correct adjustments are made in the correct order, and verifies the device is within the accuracy tolerances specified when complete.

■ OBJECTIVES

After completing this chapter, you should be able to:

- Describe the importance of the calibration procedure.
- List the calibration standards required for the calibration procedure.
- Explain the purpose of the calibration procedure.
- Give, and utilize, the formula for device accuracy.
- Give, and utilize, the formula for gain.
- Describe repeatability.
- List the requirements for standard test equipment.
- Identify a calibration error.
- Determine the required test points for a five-point calibration check.
- Record calibration data properly.
- Correct calibration errors accurately.

■ INTRODUCTION

3.1 Describing the Necessity for Calibration

We know that the operation of an automatic control process is dependent upon the accuracy of each instrument in the loop. A correctly calibrated instrument ensures the safety and proper operation of the controlled process. By definition, calibration is the process of adjusting an instrument or compiling a deviation chart so that its reading can be correlated to the actual value being measured.

3.2 Calibration Standards

For our purposes, it is essential to understand how to calibrate an instrument by using the appropriate input and output standards rather than observing the instrument's output with respect to an actual process input.

By correctly analyzing the input and output values of a transmitter using the correct standards, we can determine if necessary steps to adjust the transmitter are needed. This process, known as "bench calibration," simulates the process the instrument will measure and determines if corrective action is needed. There are times when an instrument will be calibrated after it has measured a process. Some reasons include normal maintenance and repair, regularly scheduled calibrations, and adjustments due to process deviations. In these cases, care must be taken to follow established standards for the calibration of an instrument with respect to site procedures for decontamination, calibration, etc.

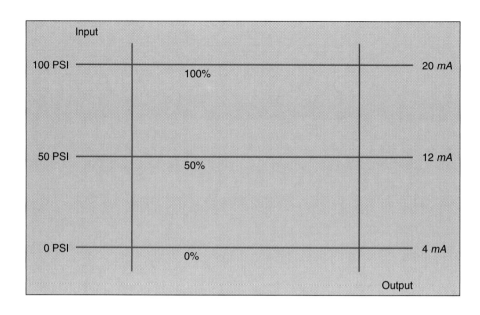

3.3 Concepts of Calibration

To properly perform an instrument calibration, it is necessary that you understand the concepts of range and span of an instrument's input and output signals. We will use a typical pressure transmitter to show our range and span. A pressure transmitter can be calibrated in pounds per square inch (PSI) and adjusted to output a corresponding signal of 4–20 mA.

From Figure 3–1, we see that our span for the transmitter is 100 PSI. The range for the same transmitter is 0 to 100 PSI. The output of the transmitter, the signal, has a span of 16 mA. This is a common output for most transmitter signals. Figure 3–1 also shows the correct relationship between the input and output of the transmitter.

When an instrument is properly calibrated, its input is a direct representation of its output, so that the:

% input is equal to its % output.

Figure 3–1 shows that at 100 PSI (100%) input, we see an output of 20 mA (100%); at 50 PSI (50%) input, we see an output of 12 mA (50%); and at 0 PSI (0%) input, we see an output of 4 mA (0%).

To properly realize the function of the correctly calibrated pressure transmitter, we need to understand that an equivalent output signal correctly represents the input signal. Every transmitter has guidelines that the device must meet to perform within the manufacturer's specifications, and every calibration performed must meet specifications for accuracy to be considered calibrated.

■ CALIBRATION PROCEDURE

3.4 Calibration Accuracy

In instrumentation, calibration must be accurate, and accuracy is a direct method of determining if an instrument must be calibrated.

This form of circular reasoning is intended to show the importance of an instrument's calibration expressed in **accuracy.** Accuracy is usually expressed as a percentage of the transmitter span. To put it simply, accuracy equals the difference between true and measured values, divided by the instrument's span, and multiplied by 100.

$$\text{Accuracy } \% = (\text{deviation/span}) \times 100$$

Let's use the pressure transmitter from Figure 3–1 as an example. Applying a 50% input to a transmitter (50 PSI) should result in a 50% output signal (12.00 mA). If an input value of 50% was applied and a measured output of 51% (12.16 mA) was recorded, we could state that the transmitter is accurate within 0.16 mA. This is one way of stating accuracy, but accuracy should always be expressed as a percent of span.

To properly understand the calculations of accuracy, we need to understand deviation. In fact, we have already calculated an instrument's deviation by comparing the difference of an instrument's input signal to its output signal. In the previous example, we saw how an applied pressure of 50 PSI (50%) was transmitted as an output signal of 12.16 mA. An input of 50% should have provided an output of 50% (12 mA). The difference, 0.16 mA, is the deviation. From this point on, use deviation to calculate accuracy:

$$\text{Accuracy } \% = (0.16 \text{ mA}/16 \text{ mA}) \times 100$$

$$\text{Accuracy } \% = 0.01 \times 100$$

$$\text{Accuracy } \% = 1\%$$

The accuracy of a transmitter can now be calculated using the formula:

$$\text{Accuracy } \% = (\text{deviation/span}) \times 100$$

Accuracy can be used to determine if an instrument must be recalibrated to meet specifications.

3.5 Gain

We now know how to calculate the accuracy of a transmitter, but the level of accuracy is dependent upon gain. **Gain** is the measure of how much an output should change given the same input change. It is the most important factor used to determine the level of accuracy. Gain equals the output signal divided by the input span:

$$\text{Gain} = \frac{\text{output signal span}}{\text{input signal span}}$$

The transmitter in Figure 3–1 has an output signal span of 16 mA and an input span of 100 PSI.

$$\text{Gain} = \frac{\text{output signal span}}{\text{input signal span}}$$

$$\text{Gain} = \frac{16 \text{ mA}}{100 \text{ PSI}}$$

$$\text{Gain} = 0.16 \; \frac{\text{mA}}{\text{PSI}}$$

Gain represents the magnitude of signal change for each PSI of input. In this example, the transmitter signal will increase or decrease to a respective input change of 1 PSI by 0.16 mA. Gain directly affects calibration. A transmitter with a low ratio of input to output has a high gain and can reflect a more noticeable change in output signal from a change in input pressure. This results in an instrument that can be more accurately calibrated.

3.6 Repeatability

With so much emphasis on the accuracy of a transmitter, it is easy to understand that the precision of an instrument is important also. A precise instrument will record an identical output signal each time an identical input is applied. To put it simply, **precision** is the **repeatability** of an instrument to record the same output signal to its corresponding input.

To clarify, accuracy is how closely an instrument reflects its input while repeatability is a term given to an instrument that reflects the same accurate output signal from a constant input signal.

3.7 Standard Test Equipment

How do we know when an instrument is correctly calibrated to a certain accuracy and precision? We use standards that establish guidelines for acceptable instrument calibration. For each input to a transmitter, an "input test standard" must be used; likewise for the output—an "output test standard" must be applied. For the transmitter in Figure 3–1, the input test standard must be capable of supplying pressure (PSI) with sufficient units of measurement to obtain the required accuracy (Figure 3–2). For the output, milliamps must be measured with enough units of **resolution** to achieve the required accuracy.

For example, if an output test standard reading milliamps was used and it could not record "tenths" of a milliamp (0.0001 A), an accuracy of 6.25% is the best accuracy measurement that can be obtained:

$$\text{Accuracy \%} = (1.0 \text{ mA}/16 \text{ mA}) \times 100$$

$$\text{Accuracy \%} = 0.0625 \times 100$$

$$\text{Accuracy \%} = 6.25\%$$

Manufacturers provide some standards to the degree of accuracy their device is capable of, but most often, the customer will provide them or inform you which standards to use to achieve acceptable calibrations.

The degree of accuracy and reliability of the calibration equipment is equally important. To prove that a calibration is valid when the de-

FIGURE 3–2 Pneumatic calibrator.

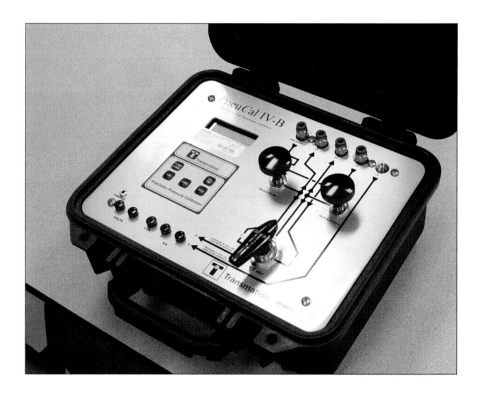

FIGURE 3–3 Certification "sticker" verifying accuracy.

sired accuracy is obtained, you should always use certified test equipment to calibrate. Certified equipment is the only assurance that a desired output is truly recorded as accurate, rather than some incorrect value due to a faulty test equipment reading. You can obtain certified test equipment by comparing and calibrating the test equipment to a certified test instrument that meets known standards (Figure 3–3).

The National Bureau of Standards can also certify test equipment for accuracy. Some organizations that promote safety and quality for instrumentation and control systems provide methods that, if followed, ensure the accuracy and performance of control systems. Such equipment is usually dedicated equipment for standardizing or verifying the accuracy of other calibration equipment and/or instruments.

3.8 Identification of Calibration Errors

Calibration is the process of adjusting or compiling a deviation chart so its reading can be correlated to the actual value being measured. We know that an instrument can be calibrated to within certain specifications of accuracy and must have sufficient repeatability to maintain accuracy. Therefore, to calibrate any instrument, the errors in the instrument must be identified.

Instrument errors are discovered by comparing the measured output to the expected output given a specific input. There are several types of instrument errors for which calibration can adjust: zero shift, span error, nonlinearity, dead band, and hysteresis are the most common.

Zero shift is the term for an instrument whose output is consistently higher or lower than the expected value. This "shift" is consistent throughout the output signal span and, to state it correctly, "the deviation is consistent throughout the signal span."

FIGURE 3–4 Recorded output
signal of transmitter in Figure 3–1.

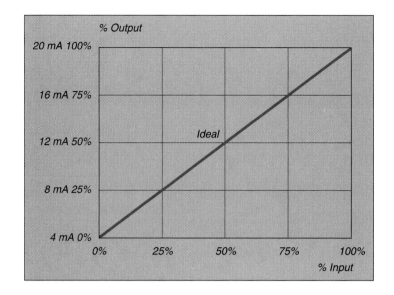

FIGURE 3–5 Transmitter from
Figure 3–1 with "zero shift" errors
recorded.

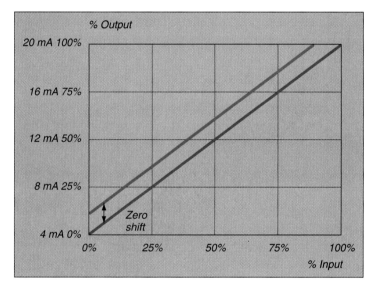

Figure 3–4 shows the expected output of our transmitter from Figure 3–1. Figure 3–5 shows the same transmitter but with "zero shift" that has elevated the output. An elevated output suggests that the output signal "starts" higher than expected. Zero shift could also "start" out lower than expected and, in this case, the zero shift would be suppressed. For either instrument error, turning the **zero adjustment** screw can correct the problem. However, it is important that the output be checked again to verify that the output is correct. Sometimes it is necessary to repeat the steps several times.

3.9 The Five-Point Check

A five-point check is performed to verify that an instrument is properly calibrated over its full output signal span. A simulated input is applied at 0%, 25%, 50%, 75%, and 100% of input range. The output is recorded for each input and the measured output is compared to the expected output. There are other suggested values to be used rather than

FIGURE 3-6 Calibration data record.

Manufacturer:	
Instrument ID:	Model:
Calibration Range:	
Input:	Output:

Test Equipment	Model	S/N

Accuracy of calibration is 0.5%

%	INPUT	DESIRED	AS FOUND	AS LEFT
0				
25				
50				
75				
100				
75				
50				
25				
0				

Performed by:_____

Date: _____

Calibration Procedure 1151 DPT

• Select and connect input source as required.
• Select and connect for voltage supply and current readout.
• Input five cardinal test points, increasing and decreasing, while recording necessary information.
• Make adjustments, if required, to comply with accuracy limits.
• Record all necessary information.
• Complete all other information as per calibration sheet.

0%, 25%, 50%, 75%, and 100%, but a calibration check that covers the full span will detect errors. In addition, the five-point check may be able to detect other errors and also provides a clearer method for detecting errors. Calibration sheets are often used to record calibration data; these

Input values		Output values	
Test points	Inputs	Expected	Actual
0%	0 PSI	4 mA	4.6 mA
25%	25 PSI	8 mA	8.6 mA
50%	50 PSI	12 mA	12.6 mA
75%	75 PSI	16 mA	16.6 mA
100%	100 PSI	20 mA	20.6 mA

FIGURE 3–7 Calibration sheet of the transmitter in Figure 3–5.

sheets can be helpful when you try to determine an instrument's calibration errors. A graph such as the one in Figure 3–4 is often used to plot the measured output signal values so they can be compared to the expected output signal values. This visual approach can help you to identify errors.

It is considered good practice to check output signal readings on an up-scale and down-scale check. Checking up-scale tests a transmitter's output signal starting at 0% and increasing to 100%. Checking a transmitter down-scale measures output signals starting at 100% and decreasing to 0%. The values recorded for both should be identical (Figure 3–5).

Figure 3–6 displays a typical calibration sheet and its respective procedure.

Figure 3–7 shows an abbreviated instrument calibration sheet. The values recorded are the values of the pressure transmitter of Figure 3–5.

The actual output signal values recorded show a consistent shift "higher" than the expected values. The plot of Figure 3–5 is the plot of the points recorded in Figure 3–7. Figure 3–5 and Figure 3–7 show a consistent shift in the expected output signal; therefore, an instrument calibration error exists and should be corrected. A single method can be used to identify calibration errors or both methods can be used together. The intent of a five-point calibration check is to provide the simplest method to identify instrument errors. We can now identify zero shift calibration errors and, employing the same technique, we can identify span errors.

3.10 Correction of Calibration Errors

Span errors are errors in an instrument's output signal that do not reflect 100% of the output signal or do not follow the input span. You can identify span errors by performing the five-point check and recording the measured output signal values. With span error, the measured outputs will vary from the expected values, but there should not be an equal "shift value" from the expected output. Figure 3–8 is the graph

FIGURE 3–8 Transmitter with span error displayed by plotting the output signal.

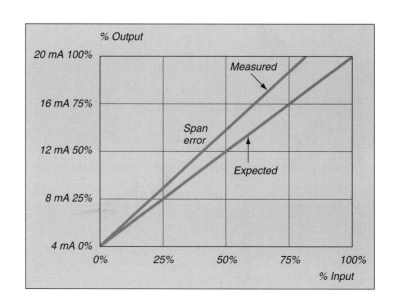

of an instrument with span error. The measured outputs begin at the origin but do not follow the expected outputs.

Instruments with span error do not produce outputs consistent with some small shift from expected outputs. If we understand the concept of deviation completely, we can define span error as deviations in an instrument's measured output that are not consistent with a "zero shift." To correct a span error in an instrument, the span adjustment can be tuned to eliminate span errors.

To properly calibrate an instrument with span errors requires a five-point check to ensure that the values of the output are accurate over the working range of the transmitter. It is important to begin calibration of a transmitter by eliminating zero shift values first and then adjusting the span adjustment to eliminate span errors. When span errors are eliminated, it is important to check for zero shift errors again because the zero setting may have shifted (due to the construction of the transmitter) when span errors were adjusted. It will be necessary to repeat the steps multiple times to achieve accurate results.

Zero and span errors are the most common errors found in an instrument's calibration. Often enough, both zero shift and span errors exist simultaneously. Figure 3–9 shows an instrument with both zero shift and span errors. To correct these problems, the instrument must first be "zeroed" and then span should be adjusted. When span is adjusted, the zero setting needs to be checked several times in order to calibrate again. The process may need to be repeated within the instrument's specifications.

Other calibration errors may exist, including nonlinearity. As you may have noticed, an instrument's output will be a linear plot that may or may not be parallel to its expected output. Nonlinearity will produce an output that, when measured and plotted, will not have any consistent error shift between the upper and lower limits of its range. It is important to realize that a large linearity problem will not be correctable and the instrument must be repaired. In some cases, nonlinearity errors are small enough that their effect is scarcely noticeable, but these errors will

FIGURE 3–9 Transmitter with span and zero errors displayed by plotting the output signal.

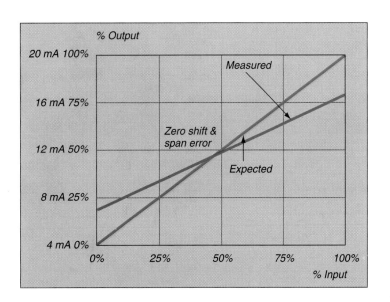

FIGURE 3–10 Transmitter with nonlinearity errors displayed by plotting the output signal.

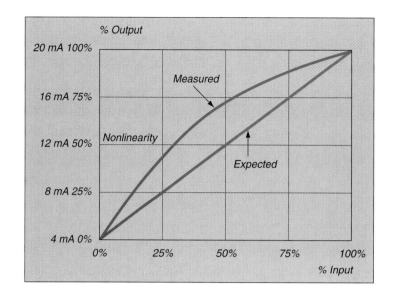

FIGURE 3–11 Transmitter with hysteresis errors displayed by plotting the output signal.

Input values		Output values		
Test points	Inputs	Expected	Upscale	Downscale
0%	0 PSI	4 mA	4.6 mA	3.4 mA
25%	25 PSI	8 mA	8.6 mA	7.4 mA
50%	50 PSI	12 mA	12.6 mA	11.4 mA
75%	75 PSI	16 mA	16.6 mA	15.4 mA
100%	100 PSI	20 mA	20.6 mA	19.4 mA

FIGURE 3–12 Rosemount 1151 Alphaline pressure transmitter "zero & span" adjustment screws.

often increase with time. You can identify nonlinearity on an instrument calibration sheet by the lower range value matching the expected value and the upper range value matching the expected value. However, the measured values between 0% and 100% do not match the expected values. Figure 3–10 is the plot of a transmitter with nonlinearity errors.

Hysteresis is another error that can be found in measuring devices. You can identify hysteresis when an instrument has measured output values that differ up-scale from down-scale. With a five-point check, hysteresis will produce a deviation from expected values when progressing up-scale from 0% to 100%. Checking the instrument down-scale from 100% to 0% will produce an equal deviation opposite from the up-sale check and only the direction from which the test point is approached will cause the error. Recording output measurements throughout the input range will cause different outputs when moving up or down. Hysteresis is not a common error found in electronic instruments—it usually occurs in mechanical measuring devices. Figure 3–11 contains data from a typical instrument with hysteresis error.

Another error that can be identified as an instrument error is **dead band.** Dead band occurs when the input value can be varied but no visible output change can be recorded. Dead band can be a very small value that is within the tolerances specified. When the dead band error becomes significant enough to stray beyond the tolerances specified, the instrument may be in need of repair. Dead band is not generally adjustable, but the error magnitude may be reduced by increasing the gain of the instrument. Increasing the gain (decreasing the span) results in an easily recorded fluctuation in the output signal. A great number of instrumentation errors can be identified through interpretation of input/output data. Calibrating an instrument means that it must be tested, recorded, and adjusted to perform as needed. If you know how to record an instrument's response, you can perform the job of calibration easily.

FIGURE 3–13 Decade box, used to simulate resistance.

FIGURE 3–14 Wallace & Tiernan portable pneumatic calibrator, "Wally Box."

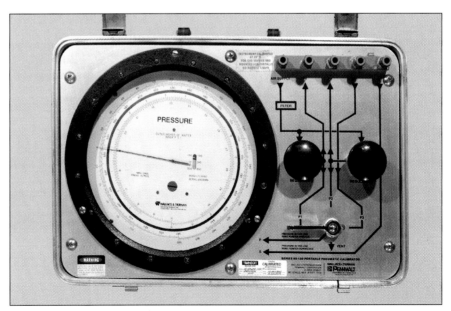

FIGURE 3–15 Yokogawa CA12 Handy Cal, calibrates RTD and thermocouples.

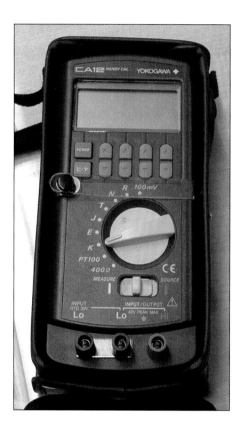

■ SUMMARY

The calibration procedure is not difficult to complete correctly, but it does require adherence to a set procedure. This chapter provided the relevant information that you will need to complete the process of calibration. Recall the text discussion of the calibration procedure and notice that "calibration equipment" was not discussed. You will need knowledge of each piece of equipment used to calibrate so that you can use it properly. The procedures described for calibrating devices in this chapter should allow you to become familiar with calibration equipment and perform the procedure correctly.

■ REVIEW QUESTIONS

1. Determine the five test points used for a temperature transmitter with calibration range of 32°F–212°F, and 4–20 mA output.

2. List the standard test equipment that would be required to calibrate a differential pressure transmitter measuring H_2O with a 4–20 mA output.

3. Determine the accuracy of a transmitter that has a maximum deviation of 0.18 mA and an output span of 16 mA.

4. What is the maximum allowed tolerance in mA for a device that has an accuracy requirement of 0.5%?

5. Determine the gain for a transmitter with input span of 20 H_2O and an output signal of 4–20 mA.

6. What calibration error is present when a device records a consistent error of 0.2 mA at each of the five test points?

7. When zero shift and span errors are identified in a device, which of the two errors is corrected first?

8. When span is adjusted on a device, it is important to check the device for what error?

chapter 4

Fundamentals of Pressure

■ OUTLINE

■ OVERVIEW

Pressure measurement can provide a measured range of variables that can be used for process control. Flow, level, and temperature measurements are all obtainable as inferred variables through the direct measurement of pressure. This chapter outlines the basic principles that apply to gauge pressure, absolute pressure, and differential pressure measurements. Obtaining a usable pressure measurement in a form that can be interpreted to an actual flow, level, or temperature is directly dependent upon the installation requirements of the pressure-sensing device. This chapter explains the operation of pressure sensors and their related applications to provide the necessary requirements for correct installation.

■ OBJECTIVES

After completing this chapter, you should be able to:

- Describe pressure measurement applications.
- List the types of elements used to measure pressure.
- Describe the functions of pressure elements.
- List the types and operation of various temperature elements given by the reference.
- List the types and operation of various flow elements given by the reference.
- List the types and operation of various level-measurement devices.
- Determine the function of pressure-sensing devices.
- Identify a pressure-operating element.

■ INTRODUCTION

4.1 Pressure Fundamentals

Pressure is the result of a force acting over a given area. Pressure can result from one object set upon another; from elevating liquids some distance above another object, such as a water tower; from the expansion of a gas; or from the force of a fluid flow.

Pressure is a universal processing condition because all forms of life depend on pressure for survival. The atmospheric pressure enables all of us to have oxygen to breathe, and to control movements, etc. Pressure supplies us with water for various uses. In a typical processing plant, pressure is responsible for the process reactions that cause the proper boiling points, condensation points, costs, and more. The measure of pressure or, in some cases, the lack of pressure (vacuum), is a critical function. Instruments can be installed to cover a wide range of pressure measurements. We next discuss how these measurements are used.

■ ELEMENT TYPES

4.2 Bellows Pressure-Sensing Element

The need for a pressure-sensing element that is extremely sensitive to low pressures and provides power for activating recording and indicating mechanisms resulted in the development of the metallic **bellows** pressure-sensing element. The metallic bellows is most accurate in measuring pressures from 0.5 to 75 PSIG. However, when used

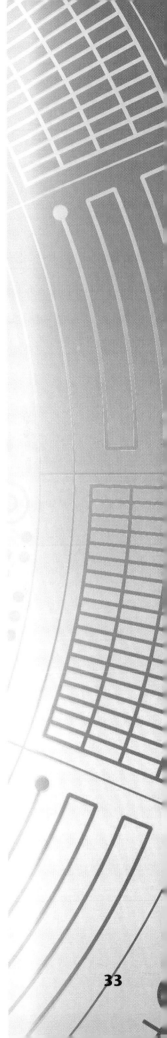

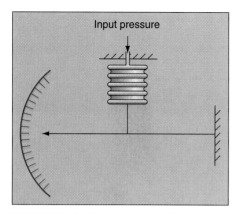

FIGURE 4–1 Basic metallic bellows.

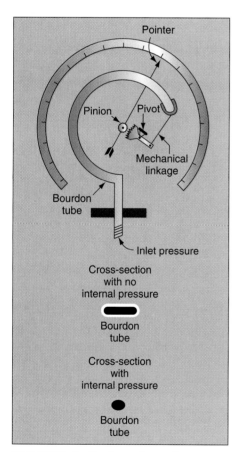

FIGURE 4–2 Bourdon tube.

in conjunction with a heavy range spring, some bellows can be used to measure pressures of over 1,000 PSIG. Figure 4–1 shows a basic metallic bellows pressure-sensing element.

The bellows is a one-piece, collapsible, seamless, metallic unit that has deep folds formed from very thin-walled tubing. The diameter of the bellows ranges from 0.5 to 12 inches and may have as many as twenty-four folds. System pressure is applied to the internal volume of the bellows. As the inlet pressure to the instrument varies, the bellows will expand or contract. The moving end of the bellows is connected to a mechanical linkage assembly. As the bellows and linkage assembly move, either an electrical signal is generated or a direct pressure indication is provided. The flexibility of a metallic bellows is similar in character to that of a helical, coiled compression spring. Up to the elastic limit of the bellows, the relation between increments of load and deflection is linear. However, this relationship exists only when the bellows is under compression. It is necessary to construct the bellows such that all of the travel occurs on the compression side of the point of equilibrium. Therefore, in practice, the bellows must always be opposed by a spring, and the deflection characteristics will be the resulting force of the spring and bellows.

4.3 Bourdon Tube-Type Detectors

The bourdon tube pressure instrument is one of the oldest pressure-sensing instruments in use today. The bourdon tube (refer to Figure 4–2) consists of a thin-walled tube that is flattened diametrically on opposite sides to produce a cross-sectional area elliptical in shape, having two long, flat sides and two short, round sides. The tube is bent lengthwise into an arc of a circle of 270° to 300°. Pressure applied to the inside of the tube causes distention of the flat sections and tends to restore its original, round cross-section. This change in cross-section causes the tube to straighten slightly. Since the tube is permanently fastened at one end, the tip of the tube traces a curve that is the result of the change in angular position with respect to the center. Within limits, the movement of the tip of the tube can then be used to position a pointer or to develop an equivalent electrical signal (discussed later in the text) to indicate the value of the applied internal pressure.

Table 4–1 summarizes the bellows and bourdon tube pressure detectors.

■ PRESSURE DETECTOR FUNCTIONS

4.4 Pressure Detector Operation

Although the pressures that are monitored vary slightly depending on the details of facility design, all pressure detectors are used to provide up to three basic functions: indication, alarm, and control. Since the fluid system may operate at both saturation and subcooled conditions, accurate pressure indication must be available to maintain proper cooling. Some pressure detectors have audible and visual alarms that sound when specified, preset limits are exceeded. Some

Table 4–1 **Summary of Bellows and Bourdon Tube Pressure Detectors**

In a Bellows-Type Detector:	In a Bourdon Tube-Type Detector:
System pressure is applied to the internal volume of a bellows and mechanical linkage assembly.	System pressure is applied to the inside of a slightly flattened, arc-shaped tube. As pressure increases, the tube tends to restore to its original, round cross-section. This change in cross-section causes the tube to straighten.
As pressure changes, the bellows and linkage assembly move to cause an electrical signal to be produced or to cause a gauge pointer to move.	Since the tube is permanently fastened at one end, the tip of the tube traces a curve that is the result of the change in angular position with respect to the center. The tip movement can then be used to position a pointer or to develop an electrical signal.

pressure-detector applications are used as inputs to protective features and control functions.

Detector Failure

If a pressure instrument fails, spare detector elements can be utilized if installed. If spare detectors are not installed, the pressure may be read at an independent local mechanical gauge, if available, or a precision pressure gauge can be installed in the system at a convenient point. If the detector is functional, it may be possible to obtain pressure readings by measuring voltage or current values across the detector leads and comparing this reading with calibration curves.

Environmental Concerns

Pressure instruments are sensitive to variations in the atmospheric pressure surrounding the detector. This is especially apparent when the detector is located within an enclosed space. Variations in the pressure surrounding the detector will cause the indicated pressure from the detector to change; this will greatly reduce the accuracy of the pressure instrument. You should consider this when you install and maintain these instruments.

Ambient temperature variations will affect the accuracy and reliability of pressure-detection instrumentation. Variations in ambient temperature can directly affect the resistance of components in the instrumentation circuitry, and, therefore, affect the calibration of electric/electronic equipment. The effects of temperature variations are reduced by the design of the circuitry and by maintaining the pressure-detection instrumentation in the proper environment.

Table 4–2   Functional Uses Summary

Pressure Detectors Perform the Following Basic Functions:	If a Pressure Detector Becomes Inoperative:	Environmental Concerns:
Indication	A spare detector element can be used (if installed).	Atmospheric pressure
Alarm	A local mechanical pressure gauge can be used (if available).	Ambient temperature
Control	A precision pressure gauge can be installed in the system.	Humidity

The presence of humidity will also affect most electrical equipment, especially electronic equipment. High humidity causes moisture to collect on the equipment, causing short circuits, grounds, and corrosion, which, in turn, may damage components. The effects due to humidity are controlled by maintaining the equipment in the proper environment.

Table 4–2 summarizes the functional uses of pressure detectors.

■ PRESSURE AND TEMPERATURE RELATIONSHIPS

There is a distinct relationship between pressure and temperature in the sense that as fluids, solids, liquids, and gases increase in pressure, they also increase in temperature. This concept is often used to measure the temperature of fluids as well as pressure. Sealed enclosures of fluids sense the change in temperature and transmit that change to pressure recorders that can be calibrated in degrees of temperature or pressure.

4.5 Resistance-Type Detectors

Resistance-Type Transducers

Included in this category of transducers are strain gauges and moving contacts (slidewire variable resistors). Figure 4–3 illustrates a simple strain gauge, which measures the external force (pressure) applied to a fine wire. The fine wire is usually arranged in the form of a grid. The pressure change causes a resistance change due to the distortion of the wire. The value of the pressure can be determined by measuring the change in resistance of the wire grid. The following equation shows the relationship to pressure and resistance:

$R = K(^L/_A)$

R = resistance of the wire grid in ohms

K = resistivity constant for the particular type of wire grid

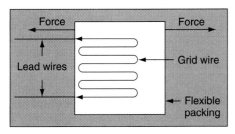

FIGURE 4–3 Strain gauge.

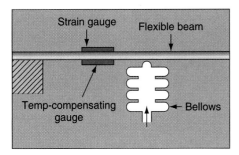

FIGURE 4–4 Strain gauge pressure transducer.

FIGURE 4–5 Strain gauge used in a bridge circuit.

L = length of wire grid

A = cross-sectional area of wire grid

As the wire grid is distorted by elastic deformation, its length is increased and its cross-sectional area decreases. These changes cause an increase in the resistance of the wire of the strain gauge. This change in resistance is used as the variable resistance in a bridge circuit that provides an electrical signal for indication of pressure. Figure 4–4 illustrates a strain gauge pressure transducer.

An increase in pressure at the inlet of the bellows causes the bellows to expand. The expansion of the bellows moves a flexible beam to which a strain gauge has been attached. The movement of the beam causes the resistance of the strain gauge to change. The temperature-compensating gauge compensates for the heat produced by current flowing through the fine wire of the strain gauge. Strain gauges, which are nothing more than resistors, are used with bridge circuits, as shown in Figure 4–5.

Alternating current is provided by an exciter that is used in place of a battery to eliminate the need for a galvanometer. When a change in resistance in the strain gauge causes an unbalanced condition, an error signal enters the amplifier and actuates the balancing motor. The balancing motor moves the slider along the slidewire, restoring the bridge to a balanced condition. The slider's position is noted on a scale marked in units of pressure.

Other resistance-type transducers combine a bellows or a bourdon tube with a variable resistor, as shown in Figure 4–6. As pressure changes, the bellows will either expand or contract. This expansion and contraction causes the attached slider to move along the slidewire, increasing or decreasing the resistance, and thereby indicating an increase or decrease in pressure.

Inductance-Type Transducers

The inductance-type transducer consists of three parts: a coil, a movable magnetic core, and a pressure-sensing element. The element is attached to the core and, as pressure varies, the element causes the core to move inside the coil. An ac voltage is applied to the coil and, as the core moves, the inductance of the coil changes. The current through the coil will increase as the inductance decreases. For increased sensitivity, the coil can be separated into two coils by utilizing a center tap,

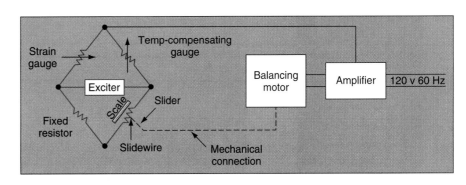

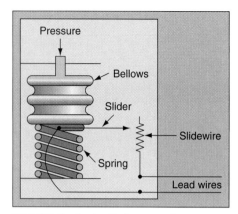

FIGURE 4–6   Bellows resistance transducer.

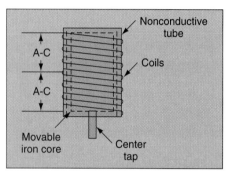

FIGURE 4–7   Inductance-type pressure transducer coil.

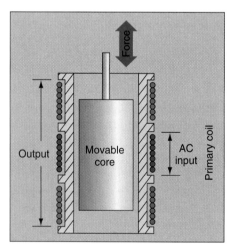

FIGURE 4–8   Differential transformer.

as shown in Figure 4–7. As the core moves within the coils, the inductance of one coil will increase, while the other will decrease.

Another type of inductance transducer, illustrated in Figure 4–8, utilizes two coils wound on a single tube and is commonly referred to as a *differential transformer.* The primary coil is wound around the center of the tube. The secondary coil is divided with one half wound around each end of the tube. Each end is wound in the opposite direction, which causes the voltages induced to oppose one another. A core, positioned by a pressure element, is movable within the tube. When the core is in the lower position, the lower half of the secondary coil provides the output. When the core is in the upper position, the upper half of the secondary coil provides the output. The magnitude and direction of the output depend on the amount the core is displaced from its center position. When the core is in the mid-position, there is no secondary output.

Capacitive-Type Transducers

Capacitive-type transducers, illustrated in Figure 4–9, consist of two flexible, conductive plates and a dielectric. In this case, the dielectric is the fluid. As pressure increases, the flexible conductive plates will move farther apart, changing the **capacitance** of the transducer. This change in capacitance is measurable and is proportional to the change in pressure.

4.6 Pressure and Flow

A fluid flowing in a line or pipe encounters restrictions in the form of couplings, fittings, flanges, etc. Special restrictions apply in recording these pressure changes with the resulting flow measurements. This concept is studied in more detail under the "Flow" heading in chapter 5.

4.7 Pressure and Level

It is important to understand how the physical properties of the measured process affect all measuring-detection devices. Devices that measure pressure can be employed for a variety of applications: pressure, temperature, flow, and level. The following provides a detailed description of the properties of level measurements utilizing pressure.

Again, pressure is the result of force acting over a given area and is often defined in terms of **head.** "Head" is defined as a column of water of certain height exerting some pressure upon some measuring point with units in PSIG. To understand "head," observe a column of water 30 feet tall sitting on a pressure sensor (Figure 4–10). The weight of the column can be derived by first finding the volume of the column. This is the area of the base times the height of the liquid, or 1 times 30 equals 30 cubic feet of water.

The approximate weight of water is 62.43 pounds per cubic foot ($^{lb}/_{ft}{}^3$). Therefore, the total weight of the column will equal 30 ft $\times$ 62.43 $^{lb}/_{ft}{}^3 = 1870.2$ $^{lb}/_{ft}{}^2$. Pressure is normally recorded in "inches" (PSI) rather than feet, so there are 144 square inches per square foot

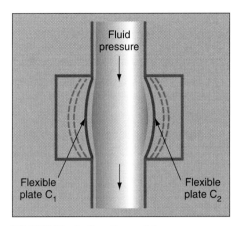

FIGURE 4-9 Capacitive pressure transducer.

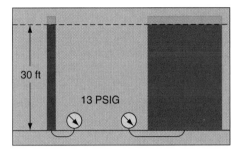

FIGURE 4-10 Liquid head (pressure) measurement resulting from mass.

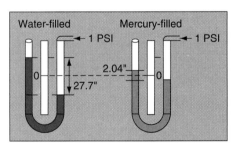

FIGURE 4-11 Specific gravity relationships between water and mercury.

(in²/ft²) and if we divide 1870.2 $^{lb}/_{ft^2}$ by 144 $^{in^2}/_{ft^2}$ we get 12.9875 lb/in², or roughly 13 PSIG.

$$30 \text{ ft} \times 62.43 \frac{\text{lb}}{\text{ft}^3} = 1870.2 \frac{\text{lb}}{\text{ft}^2}$$

Therefore:

$$\frac{1870.2 \dfrac{\text{lb}}{\text{ft}^2}}{144 \dfrac{\text{in}^2}{\text{ft}^2}} = 12.9875 \frac{\text{lb}}{\text{in}^2} \text{ or PSI}$$

Fluid levels can also be calculated using the pressure sensor because all fluids create pressure in direct relationship with their **density** and depth. As the depth of a pressure increases, the pressure of the fluid also increases. As stated previously, the head of a fluid can be used to calculate the resulting pressure. To measure pressure and its related value, we must first understand to what we are relating our pressure in order to understand its true indication. We know that a cubic foot of water weighs 62.43 lb/ft³. Suppose that same cubic foot (volume) contains mercury instead. Mercury has a specific gravity that is 13.6 times heavier than water, so the resulting weight of the mercury will be approximately 850 lb.

$$62.43 \frac{\text{lb}}{\text{ft}^3} \times 13.6 = 849.048 \frac{\text{lb of mercury}}{\text{ft}^3}$$

To fully understand this is to realize that all fluids have different weights due to their densities. To keep a uniform sense of all pressure measurements, their densities are referenced to water to derive a ratio value known as **specific gravity.**

Specific Gravity

Figure 4–11 shows two manometers: one filled with water and the other with mercury. Both have an applied pressure of 1 PSI. The water manometer is indicating a pressure difference of 27.7 inches, while the mercury manometer indicates a difference of 2.04 inches. With mercury being 13.6 times heavier than water, it follows that an identical pressure will move the mercury less. Using mercury, the range of pressure measurements can be expanded up to 13.6 times an equivalent measurement using water.

While the measurement of pressure with a manometer is a useful tool, a manometer can practically measure up to 100 inches of water column, which is the equivalent measurement of 3.6 PSI. While this range is useful, it is apparent that pressure measurements greater than 3.6 PSI will be needed. By using liquids that have a specific gravity greater than that of water, we can extend the range of the manometer. In the previous example using mercury, we can extend our range of measurement up to 13.6 times what we could measure with water. Now our maximum pressure measurement will be 13.6 times 3.6 PSI, for a result of approximately 49 PSI, clearly a more usable range. Almost all

Table 4–3   Gravity Numbers for the Most-Used Substances

Liquids		Gases	
Water	1.00	Air	1.00
Mercury (Hg)	13.6	Hydrogen	0.06
Alcohol	0.79	Nitrogen	0.96
Gasoline	0.67	Oxygen	1.10

substances have been assigned specific gravity numbers, but each substance is referenced to a substance in its own group, such as a gas to air, or a liquid to water. Table 4–3 lists the specific gravity numbers for the most-used substances.

In Figure 4–12, we see a tank filled to a level of 10 feet with alcohol. It has a pressure gauge to read the pressure of the liquid. The specific gravity of alcohol is 0.79, so if we find an equivalent water-filled tank we can use the specific gravity value to find the specific gravity of the alcohol. Water has a weight of 62.43 lb/ft³, so if we multiply 62.43 lb/ft³ times the 10 feet of liquid height, we get a result of 624.3 lb/ft². Now recall that the most common measurement for pressure is in inches, so divide 624.3 lb/ft² by 144 in²/ft². The result is 4.34 lb/in². Now multiply the specific gravity of alcohol times the pressure of water. The reading for the equivalent amount of alcohol is 3.4 pounds per square inch.

The following shows the mathematical approach:

$$62.43 \ (\text{lb/ft}^3) \times 10 \ \text{ft} = 624.3 \ (\text{lb/ft}^2)$$

$$\frac{624.3 \ (\text{lb/ft}^2)}{144 \ (\text{in}^2/\text{ft}^2)} = 4.34 \ (\text{lb/ft}^2)$$

$$4.34 \ (^{\text{lb}}/_{\text{in}}{}^2) \times 0.79 = 3.43 \ (^{\text{lb}}/_{\text{in}}{}^2)$$

FIGURE 4–12 Liquid "head" measurement displayed as pressure.

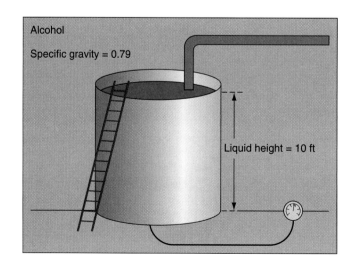

Alcohol

Specific gravity = 0.79

Liquid height = 10 ft

You can use this method on any liquid as long as you know its specific gravity. It is important to understand that the relationship of specific gravity allows us to be able to measure the pressure of any substance with a common calibration tool. Our pressure unit of measurement has been in the form of pounds per square inch, but we need to learn the other forms of pressure measurements. Pressure is measured in one of three different scales, but we will see that all three of these scales are related to each other and that knowing one, we know the other two.

Gauge Pressure

The measurement of **gauge pressure** is extremely useful in industry because this reading can show the actual force being applied to some object such as a vessel or a line. Gauge pressure is really a special form of **differential pressure**—gauge pressure readings are pressure readings without the normal atmospheric pressure included. We are all familiar with atmospheric pressure, but maybe without its fullest meaning. We know that oxygen is heavier (specific gravity = 1.10) than air (specific gravity = 1.00), therefore the closer to sea level, the higher the atmospheric pressure. As humans, we do not feel the result of atmospheric pressure because the pressure inside of us equals the outside pressure applied. For gauges that record gauge pressure, this is also true. Gauge pressure starts at 14.7 lb absolute pressure and measures increasing amounts of applied pressure. A reading of 10 PSIG is in excess of atmospheric pressure. A reading of less than atmospheric pressure records is in vacuum, or is less than atmospheric pressure. It is common practice to give gauge pressure readings (PSIG) without specifying absolute or differential or just in the form PSI. If any confusion should arise, specifying the pressure as gauge pressure should eliminate it (Figure 4–13).

Absolute Pressure

Absolute pressure is measured from the point of zero atmospheric pressure (approximately −14.7 PSIG). Since atmospheric pressure starts at total vacuum, a PSIA gauge lying on a work table at sea level would indicate 14.7 PSIA while a PSIG gauge would indicate 0. Absolute pressure gauges are less common because they require a vacuum chamber. This makes the gauge more expensive to manufacture and use (Figure 4–14).

Vacuum Pressure

Vacuum pressure is any pressure below atmospheric pressure and, as reference, gauge pressure is any point above atmospheric pressure. When there is any doubt whether a pressure is gauge (G), absolute, or differential, the pressure should be indicated in full. However, it is becoming common practice to specify gauge pressure without specifying "G" and to say "absolute" or "differential" only when each applies. Differential pressure can be used to specify vacuum or gauge by indicating whether a reading is positive or negative (Figure 4–15).

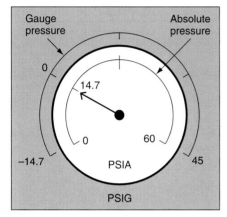

FIGURE 4–13 Absolute pressure and gauge pressure.

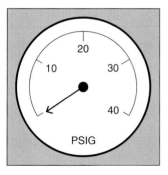

FIGURE 4–14 Gauge pressure.

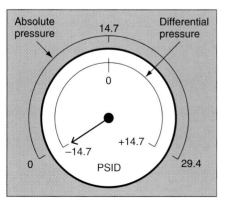

FIGURE 4–15 Absolute pressure and differential pressure.

Sometimes pressure switches are used to indicate process variables for alarm and control (Figure 4–16). For example, a pressure switch with Tag Number PSL is set to trip at 20 PSI. To calibrate the switch for the correct setpoint trip, pressure needs to be applied in excess of the trip setting and decreased to the setpoint because the tag indicates a pressure switch low setpoint, which tells us the normal operating pressure(s) are above the trip setting (Figure 4–17). Therefore, to set the

FIGURE 4–16 Adjustable pressure switch sends alarm when tripped.

FIGURE 4–17 Gauge, absolute, and differential pressure relationships.

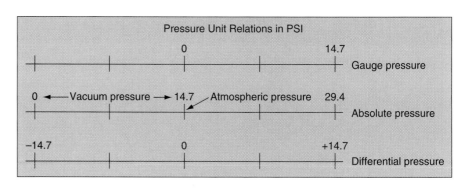

FIGURE 4–18 Pressure conversions.

Pressure Conversion Chart						
PSI	in of H$_2$O	in of Hg	mm of H$_2$O	mm of Hg	bar	mbar
.01	.2768	.0204	7.031	.5171	.0007	.6895
.1	2.768	.2036	70.31	5.171	.0069	6.895
.433	12	.8822	304.65	22.41	.0299	29.88
1	27.68	2.306	703.1	51.71	.0690	68.95
3	83.04	6.108	2109	155.1	.2068	206.8
6	166.1	12.22	4218	310.3	.4137	413.7
10	276.8	20.66	7031	517.1	.6895	689.5
14.7	406.9	29.93	10340	760.2	1.104	1014
15	415.2	30.54	10550	775.7	1.034	1034
30	830.4	61.08	21090	1551	2.068	2068
50	1384	101.8	35150	2586	3.447	3447
100	2768	203.6	70310	5171	6.895	6895
150	4152	305.4	105500	7757	10.34	10340
200	5538	407.2	140600	10340	13.79	13790
250	6920	509.0	175800	12930	17.24	17240

FIGURE 4–19 Pressure vs. altitude conversions.

Pressure vs. Altitude			
Altitude (feet)	Pressure		
	in of Hg	mm of Hg	PSI
-1,000	31.02	787.9	15.25
-500	30.47	773.8	14.94
0	29.921	760.0	14.70
500	29.38	746.4	14.43
1,000	28.86	732.9	14.18
1,500	28.33	719.7	13.90
2,000	27.82	706.6	13.67
2,500	27.31	693.8	13.41
3,000	26.81	681.1	13.19
3,500	26.32	668.6	12.92
4,000	25.84	656.3	12.70
4,500	25.36	644.2	12.45
5,000	24.89	632.3	12.23
10,000	20.58	522.6	10.10
15,000	16.88	428.8	8.28
20,000	13.75	349.1	6.75
30,000	8.88	225.6	4.36
40,000	5.54	140.7	2.72
50,000	3.426	87.30	1.689
60,000	2.132	54.15	1.048
70,000	1.322	33.59	0.649
80,000	0.820	20.83	0.403

FIGURE 4–20 Gauge pressure transmitter.

correct setpoint for "trip," the pressure has to be in excess of the trip setting and decreased to the correct setting. The converse is true for a "high" pressure trip setting. Pressure needs to be increased to the correct trip setting before adjustments can be made.

FIGURE 4–21 Adjustable pressure switch.

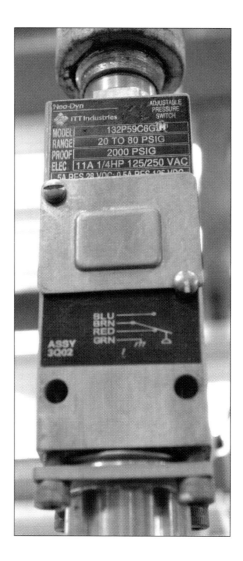

■ SUMMARY

Pressure measurement and corresponding relationships to flow, level, and temperature allow for a wide range of applications in the process control industry.

Understanding how pressure can be used to infer different measured applications is dependent upon the physical relationships described in this chapter.

■ REVIEW QUESTIONS

1. The need for a pressure-sensing element that is extremely sensitive to low pressures is a characteristic of what type of pressure-sensing device?

2. How does the bourdon tube assembly detect and display measured pressure?

3. What physical relationships provide temperature measurements from a pressure-sensing device?

4. How many inches of head is equivalent to 1 PSI?

5. What is the weight of a cubic foot of water at STP (standard test practice)?

6. Define *specific gravity*.

7. How many square inches are in a square foot?

8. How many cubic inches are in a cubic foot?

9. A pressure-sensing device that begins to measure increasing pressure from 0.0 PSIA is called what?

10. Is a pressure switch tagged PSL 1234 set to trip on increasing or decreasing pressure?

Pressure Conversion Table

Multiply no. of — To obtain →

by \ To obtain	Atmos	Bars	Dynes/Cm²	In of Hg (0°C)	In of H₂O (4°C)	K Grams/Meter²	Lb/In² PSI	Lb/Ft²	Mm of Hg torr	Microns	Pascals
Atmos	1	9.86923×10^{-1}	9.86923×10^{-7}	3.34207×10^{-2}	2.458×10^{-3}	9.678×10^{-5}	.068046	4.7254×10^{-4}	1.316×10^{-3}	1.316×10^{-6}	9.869×10^{-6}
Bars	1.01325	1	$\times 10^{-6}$	3.3864×10^{-2}	2.491×10^{-3}	9.8067×10^{-5}	6.8948×10^{-2}	4.788×10^{-4}	1.333×10^{-3}	1.333×10^{-6}	$\times 10^{-5}$
Dynes/Cm²	1.01325×10^6	10^6	1	3.386×10^4	2.491×10^3	98.067	6.8948×10^4	478.8	1.333×10^3	1.333	10
In of Hg (0°C)	29.9213	29.53	2.953×10^{-5}	1	7.355×10^{-2}	2.896×10^{-3}	2.036	.014139	3.937×10^{-2}	3.937×10^{-5}	2.953×10^{-4}
In of H₂O (4°C)	406.8	401.48	4.0148×10^{-4}	13.60	1	3.937×10^{-2}	27.68	0.1922	0.5354	5.354×10^{-4}	4.014×10^{-3}
K Grams/Meter²	1.03323×10^4	1.0197×10^4	1.0197×10^{-2}	345.3	25.40	1	7.0306×10^2	4.882	13.59	13.59×10^{-3}	1.109×10^{-1}
Lb/in² PSI	14.6956	14.504	1.4504×10^{-5}	0.4912	3.6126×10^{-2}	1.423×10^{-3}	1	6.9444×10^{-3}	1.934×10^{-2}	1.934×10^{-5}	1450×10^{-4}
Lb/ft²	2116.22	2088.5	2.0855×10^{-3}	70.726	5.202	0.2048	144.0	1	2.7844	2.7844×10^{-3}	2.089×10^{-2}
Mm of Hg torr	760	750.06	7.5006×10^{-4}	25.400	1.868	7.3558×10^{-2}	51.715	0.35913	1	$\times 10^{-3}$	7.502×10^{-3}
Microns	760×10^3	750.06×10^3	0.75006	2.54×10^4	1.868×10^3	73.558	51.715×10^3	359.1	$\times 10^3$	1	7.502
Pascals	1.01325×10^5	$\times 10^5$	$\times 10^{-1}$	3.386×10^3	2.491×10^2	9.8067	6.8948×10^3	4.788×10^1	1.333×10^2	1.333×10^{-1}	1

FIGURE 4-22 Pressure conversion table.

chapter 5

Fundamentals of Flow

OUTLINE

■ OVERVIEW

Flow measurement is probably one of the most common process variables measured today. There are many types of meters and flow characteristics that have to be understood to correctly install, set up, and calibrate. This chapter provides a detailed explanation of differential pressure, mass flow, and velocity meter applications for the process industry and discusses the operating characteristics of flow. As in the previous chapter on pressure fundamentals, not every device is examined, but if you apply the information discussed here, you can use and understand all devices for measuring flow.

■ OBJECTIVES

After completing this chapter, you should be able to:

- Describe flow measurement properties and measurement principles.
- Show flow relationships mathematically.
- Describe how differential pressure can be used to derive a flow rate.
- Explain the characteristics of flow meters.
- List two types of velocity meters.
- Explain how area flow meters operate to measure a flow.
- Describe the properties of steam measurement.
- Determine appropriate flow variables given area and mass flow rates.

■ INTRODUCTION

5.1 Flow Fundamentals

One of the basic control functions required of instrumentation is the measurement of **flow.** The control of flow is vitally important because many process variables are often governed by a regulated flow (Figure 5–1). In some operations, the ability to conduct accurate flow measurements is so important that it can make the difference between an efficient process and one that is inefficient. In other cases, the inability to make accurate flow measurements can result in disastrous results for people and equipment. Understanding the principles of flow is a basic fundamental of process measurement and control. *Fluids* are defined as liquids, gases, or vapors. For the majority of process systems, a rate of flow must be controlled, not the total flow over a set time period. The **flow rate** is often measured as a given quantity moving past a given point in a specified time period. Examples of common flow rates are **GPM** (gallons per minute) and **GPH** (gallons per hour).

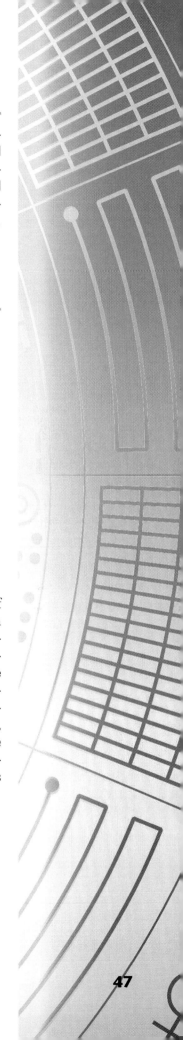

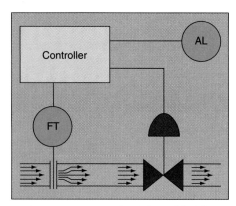

FIGURE 5–1 Principles of fluid motion.

FIGURE 5–2 Bernoulli's equation in simplified form.

■ BERNOULLI'S PRINCIPLE

5.2 Mathematical Relationships

Rate of flow can also be expressed mathematically by the equation $Q = VA$, where Q = rate of flow, V = velocity of flow (inches per second, in/sec), and A = cross-sectional (square inches, in²) area of the process line. Applying this equation, $Q = VA$, we can correctly calculate that the rate of flow (in³/second) is directly related to the velocity of the flow and the diameter of the process line. Another important fundamental that should not be overlooked is that a moving fluid contains energy. Energy in a flow consists of the sum of the pressure energy and the velocity energy. The concept was discovered by Jacques Bernoulli in the eighteenth century. Bernoulli's principle simply states that as the velocity in a given line is increased, the pressure must decrease and, conversely, when the velocity in a given line is decreased, the pressure must increase (Figure 5–2). By examining what we now know, we can make some general observations about the fundamentals of flow.

What happens to a fluid flowing in a process line when the diameter decreases and then increases? Of first importance is to realize that the flow rate, Q, will not change at any point in the process line. When the cross-sectional area (A) of the process line decreases, the velocity (V) must increase; when the cross-sectional area of the process line increases, the velocity must decrease (Figure 5–3). From Figure 5–3 we can summarize that:

$$Q = V1A1 = V2A2 = V3A3$$

5.3 Differential Pressure vs. Cross-Section

It is perhaps easier to visualize the differential pressure change through a restriction by placing a manometer across the restriction in the line and observing the effects of the process pressure upon it (Figure 5–4).

The total energy at any point in a process line will not change because energy cannot be created or destroyed. The velocity at $V2$ is greater than that at $V1$, since energy remains the same; the pressure at $P2$ is less than the pressure at $P1$. By the same reasoning, the velocity at $V3$ is less than that at $V2$ and the pressure $P3$ is greater than the pressure at $P2$. The result is that the energy is shown to balance at all points in the line, as it must.

FIGURE 5–3 Bernoulli's principle illustrated.

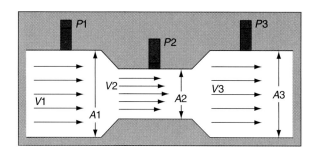

FIGURE 5–4 An example of a head flow meter.

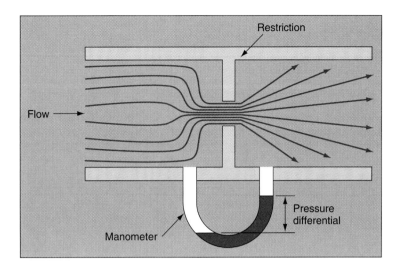

5.4 Velocity and Differential Pressure

As stated previously the flow rate, Q, is directly related to the velocity, V, by $Q = VA$. The change in pressure is not directly related but is proportional to the square of the velocity change (V). From this observation we can deduct that the velocity is equal to the square root of the pressure change, most commonly called the differential pressure, h:

$$Q = \sqrt{h} \times A$$

From this equation, we can now understand the principle of how flow measurement is obtained from differential pressure. As we measure the differential pressure in a line across a known restriction in a flow line (most commonly an orifice plate), and if we know the diameter of the line, we can calculate the actual flow rate in the line. With most liquid flow measurement instruments, the flow rate is determined by measuring the liquid's velocity or kinetic energy.

Everything discussed so far has been under "ideal" conditions, but there are other conditions that may cause some variance (Figure 5–5).

FIGURE 5–5 Venturi stream flow measurement.

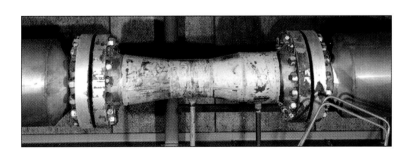

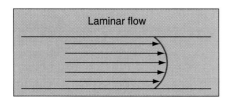

FIGURE 5–6 Laminar flow.

5.5 Laminar Flow and Turbulent Flow

Other factors that affect the flow of a fluid include the fluid's viscosity and density, and the friction of the fluid that is in contact with the inner walls of the process line. Keeping the "other" conditions in mind, we will observe that the flow in a pipe is not ideal, but, in comparatively slow rates of flow, we will see how the flow is actually laminar—the flow of a liquid is slower along the process line's inner walls than at the center of the line (Figure 5–6).

The friction that causes laminar flow can be overcome by increasing the velocity of the flow. The amount of the velocity increase depends on the viscosity of the fluid along with the diameter of the pipe. To understand this, picture a small process line (½ inch) with a flow; understand that a larger percentage of the total flow will be in contact with the inner process line walls than, say, an 8-inch line. A laminar flow is useful in that accurate flow measurements can be recorded with a wider variety of measuring instruments.

A British scientist, Sir Osborne Reynold, determined a relationship for the three basic factors that affect flow: the velocity of the fluid, the viscosity of the fluid, and the diameter of the pipe. This relationship is expressed by an equation known as Reynold's number:

$$\frac{3160 \times Q \times G_T}{D \times u}$$

Where:

Q = flow rate in GPM
G_T = specific gravity
D = inside diameter of pipe
u = viscosity of fluid (cp)

The Reynold's number equation is merely a ratio of the fluid's inertial forces (flow rate and specific gravity) to its drag forces (line diameter and viscosity). For any particular line, the inside diameter and specific gravity remain constant; therefore, it is the fluid's flow rate and viscosity that determine the type of flow for a specific pipe.

A value of Reynold's number is not critical once certain requirements are met. A Reynold's number of 3,000 or less generally indicates a laminar flow, while a Reynold's number of 5,000 or more indicates a mostly turbulent flow (Figure 5–7). Turbulent flow occurs when a mixing of flow breaks up into eddies that flow through the pipe with the same average velocity. Fluid velocity is less significant and the velocity profile is more uniform, as shown in Figure 5–7. For our purposes, it is insufficient to simply know what Reynold's number implies to a flow rate. Our basic equation for flow, $Q = VA$, can now be modified to more accurately reflect real flow rates rather than ideal flow rates. $Q = VAk$ can now be used, where k is some constant involving Reynold's number. A constant is used because the velocity is not the same at all points in a line due to some pressure loss at constricting points in a line. It is easy for us to realize what would happen if two solid surfaces were rubbed together vigorously: heat would be generated as a reflec-

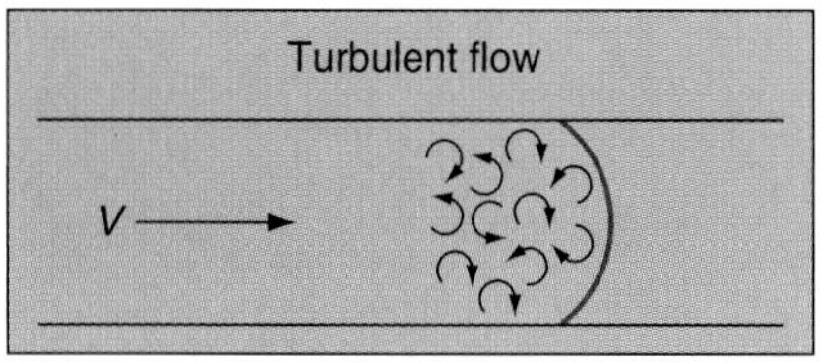

FIGURE 5–7 Turbulent flow characteristics.

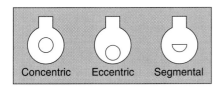

FIGURE 5–8 Orifice plates used for creating a restriction for differential head flow measurement.

tion of the energy used to cause the friction. When a fluid comes into contact with a solid surface, the same principle takes effect—energy is dissipated in the form of pressure loss (Figure 5–8). Bernoulli's principle states that the sum of the pressure energy and the velocity energy must equal the total energy in a line; therefore, a loss in pressure energy means that the velocity energy must make up the difference.

In summary:

- The rate of flow is equal to the cross-sectional area of the pipe (A), times the velocity of the fluid (V).
- The velocity (V) is equal to the square root [of differential pressure (H)].
- At relatively low values for Reynold's number, the flow is laminar.
- At relatively high values for Reynold's number, the flow is turbulent.
- Reynold's number is related to flow velocity, flow viscosity, and pipe diameter.
- The rate of flow is equal at all parts in a process line.
- A restriction in a pipe causes a permanent pressure loss and a change in pressure and velocity.
- Fluid friction is a force that opposes the flow of fluids ensuing from the presence of the fluid's viscosity and the turbulence resulting.

Thus far, we have discussed how differential pressure may exist in a process line. It is estimated that 90% to 95% of all flow devices located in the field use differential pressure to derive their flow rate; such is the reason for an in-depth study of flow and pressure characteristics.

One of the many methods used for measuring the rate of fluid flow is using the principle of pressure differential through a restricted opening. Remember that the pressure drop across a restriction is proportional to the square of the velocity or, simply figured, velocity, $V = \sqrt{(H)}$, differential pressure. By using Figure 5–1, we can see that when a fluid moves through an orifice plate, the fluid forms a concentrated flow with the lowest pressure and area smaller than the orifice plate diameter (Figure 5–9). In a line, therefore, a loss in pressure energy means that the velocity energy must make up the difference.

7-33

FIGURE 5–9 Concentrated flow at point called **vena contracta.**

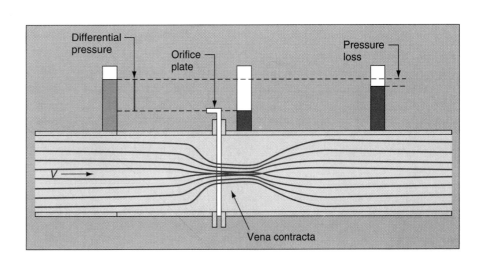

In summary:

- The rate of flow is equal to the cross-sectional area times the square root of the differential pressure. Notice the pressure eventually recovers downstream of the orifice plate but does not reach the value previously attained due to energy loss.

In practice:

- The constant, k, that is present in process flow measurement represents several factors that limit the flow rate in the process line. The result is that k can be thought of and used as a limiting factor for the flow as well. It allows for the substitution of k in mathematics to provide a useful equation for those working in the field.

$$Q = V \times A = \sqrt{H} \times A \qquad \text{(Bernoulli's principle)}$$

$$Q = \sqrt{H} \times A \times k \qquad \text{(Bernoulli's principle applied)}$$

By using the limiting factors of A (cross-sectional area) and k (friction, head loss, heat, etc.), we can substitute for these factors with the maximum flow rate (Q_{max}). In reality, A and k do limit the actual flow rate and can be substituted for. Therefore, we arrive at a new equation:

$$Q = \sqrt{H} \times A \times k$$

$$Q_{act} = \sqrt{H} \times Q_{max}$$

By applying algebra, we can derive an equation that is represented by a percentage ratio for both sides of the equation:

$$Q_{act} = \sqrt{H} \times Q_{max}$$

$$\frac{Q_{act}}{Q_{max}} = \sqrt{\frac{H_{act}}{H_{max}}}$$

By representing this equation with a percentage, we can show the percent of flow rate as being equal to a ratio that is also a percentage. This allows the substitution of the measured variable or the signal variable to be applied within the equation:

$$\%FR \text{ (flow rate)} = \sqrt{\frac{H_{act}}{H_{max}}} \quad \text{also,} \quad \%FR \text{ (flow rate)} = \sqrt{\frac{mA}{16 \text{ mA}}}$$

$$\text{Therefore, } \%FR = \sqrt{\frac{mA}{16 \text{ mA}}}$$

$$\%FR = \sqrt{\frac{mA}{16 \text{ mA}}} \text{ (substituting mA for differential pressure ratio)}$$

$$\{\%FR\}^2 = \left\{ \sqrt{\frac{mA}{16 \text{ mA}}} \right\}^2 \text{ (squaring both sides)}$$

$$\{\%Q\}^2 = \frac{mA}{16 \text{ mA}}$$

The equation can be derived:

$$\{\%Q\}^2 \times 16 \text{ mA} = \text{mA}$$

This equation allows for calculating the milliamp signal for any given process flow rate. This equation, derived from Bernoulli's principle, takes into account the square-root factor, which is present during a differential pressure flow measurement.

From previous chapters, we have determined that for a linear device, the output of the device can be calculated to provide the five-point calibration check that is needed. Specifically:

$$0\% = 4 \text{ mA}$$
$$25\% = 8 \text{ mA}$$
$$50\% = 12 \text{ mA}$$
$$75\% = 16 \text{ mA}$$
$$100\% = 20 \text{ mA}$$

Applying the previous equation gives:

$$0\% = 4 \text{ mA}$$
$$25\% = 5 \text{ mA}$$
$$50\% = 8 \text{ mA}$$
$$75\% = 13 \text{ mA}$$
$$100\% = 20 \text{ mA}$$

Therefore, for a device that has a square-root output signal, the five-point check values given above are to be used. These values should also be applied when calibrating a square-root extractor, which is used to "extract" the square root from a signal.

ELEMENTS OF FLOW METERS

5.6 Flow Characteristics

Elements of Flow Meters and Relevant Process Variables

Process measurement and control demands accuracy and dependability while interacting with various process control variables that have to be considered in each instance. The monitoring and control of a flow loop, for example, may be contingent upon the line size and velocity, but the presence of other conditions can have a direct influence on accuracy, repeatability, and dependability. Conditions such as operating temperature, ambient pressures, process reactions (exothermic or endothermic), line surges, and many other conditions may all have an effect such that the measuring device no longer performs as designed.

Even though these factors are design considerations for engineers, field technicians are confronted daily with such conditions, and many more, that will affect the functioning of their equipment. Since adverse conditions will always affect the technician's equipment, it seems relevant that the technician should be fundamentally aware of instrument selection as well as installation. Many process problems are directly related to the measuring equipment selected, installed, or implemented in a control loop. Certainly, a basic understanding of a device's advantages, or disadvantages, will be helpful in diagnosing and implementing the problem. Those who understand the requirements of the device appreciate the importance of **flow meter** selection. Engineers should recognize that each application is not the same and must be evaluated separately to determine the importance of accuracy, dependability, cost, etc. Once these factors are determined, physical factors come into play, such as:

1. Line size;
2. Range of flow rate (minimum, maximum, norm);
3. Fluid characteristics (liquid, slurry, operating pressure, operating temperature, etc.); and
4. Corrosive effects, steady or surging flow, and other conditions.

The process industry, by its very nature, deals with the rate of flow for process reactions, storage, and profit. It is estimated that flow measurement is the most common measurement performed in process plants, yet some experts claim that over 75% of the flow-measuring devices installed today are not performing satisfactorily (Figure 5–10). Faulty specifications by the designers account for some of the problems, but incorrect installation accounts for an additional amount. Field technicians who understand that the most important qualification is the flow instrument's *function* will substantially outperform their fellow workers and will contribute valuable information toward a profitable enterprise.

Although suppliers are always ready to provide information on the installation of their flow meters, estimates are that 75% of installations are performed by the users. Installation mistakes occur, and the most common is not allowing sufficient upstream and down-

FIGURE 5–10 Micro Motion mass flow meter.

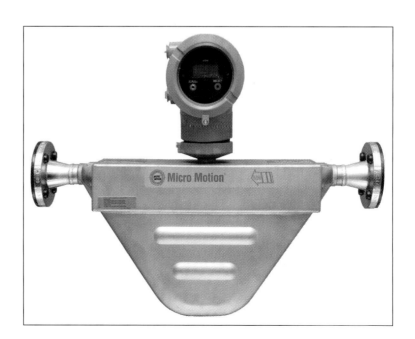

stream piping for an accurate flow measurement. With electrical components, intrinsic safety is an important consideration in hazardous locations. Most suppliers offer intrinsically safe design literature for application. Magnetic fields exist in most locations. Power lines, relays, solenoids, transformers, and motors, contribute to interference. Field technicians come into the most frequent contact with these elements and should understand their properties. Problems occur mainly with the secondary elements of devices that must be protected. Manufacturers' recommended installation instructions will usually negate these noise-contributing elements.

5.7 Flow Meter Characteristics

Space limitations prevent a detailed study of the characteristics of the various devices, so this section presents a summary. Technicians can research further if they desire. Service conditions have the greatest effect on measurement devices and numerous types of flow meters are available for service today. In general, the meters can be classified as differential pressure, positive displacement, velocity, and mass meters. The following outline shows the possibilities of flow meter selection:

    I. Displacement Meters
        A. Positive Displacement Meters
        B. Metering Pumps
    II. Differential Pressure Devices (Head Meters)
        A. Closed Measurement (Line, Vessel)
      1. Orifice
      2. Venturi Tubes
      3. Flow Tubes
      4. Flow Nozzles
      5. Pitot Tubes
  III. Other Flow Meters

Displacement Meters

Operation of displacement meters involves separating liquids into accurate, measured increments and then moving them along. Each segment is connected to a register that counts each segment as a volume amount. These units are popular with automatic batch processes and accounting applications. The meters are particularly good for applications where the measurement of viscous liquids using a simple mechanical meter system is needed.

In a displacement flow meter, all of the fluid passes through the meter in almost completely isolated quantities. The number of these quantities is counted and indicated in terms of volume or weight units by a register.

Nutating Disc

The nutating disc, or wobble plate meter, is the most common type of displacement flow meter. A typical nutating disc is shown in Figure 5–11.

FIGURE 5–11 Nutating disc meter.

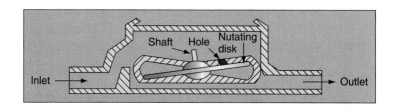

This type of flow meter is normally used for water service, such as raw water supply and evaporator feed. The movable element is a circular disc that is attached to a central ball. A shaft is fastened to the ball and held in an inclined position by a cam or roller. The disc is mounted in a chamber that has spherical side walls and conical top and bottom surfaces. The fluid enters an opening in the spherical wall on one side of the partition and leaves through the other side. As the fluid flows through the chamber, the disc wobbles, or executes a nutating motion. Since the volume of fluid required to make the disc complete one revolution is known, the total flow through a nutating disc can be calculated by multiplying the number of disc rotations by the known volume of fluid. To measure this flow, the motion of the shaft generates a cone with the point, or apex, down. The top of the shaft operates a revolution counter, through a crank and set of gears, that is calibrated to indicate total system flow. A variety of accessories, such as automatic count resetting devices, can be added to the fundamental mechanism to perform functions in addition to measuring the flow.

■ DIFFERENTIAL PRESSURE METERS

We have studied the use of differential pressure flow meters previously and we know that we can determine a liquid's rate of flow. The basic principle to understand is that the pressure drop across the meter is proportional to the square of the flow rate. The differential pressure is measured and the square root is extracted.

Differential pressure flow meters, like most other flow meters, have primary and secondary elements. The primary element is responsible for causing a change in the kinetic energy, which causes the pressure drop across the element. This unit is precisely sized for correct line size, flow rate, and liquid properties while allowing an accurate measurement over a reasonable range. The secondary element analyzes the primary element's information and provides a signal or readout that is converted to the actual flow rate.

The **orifice plate** is the simplest of the flow path restrictions used in flow detection, as well as the most economical. Orifice plates are flat plates that are 1/16- to 1/4-inch thick. They are normally mounted between a pair of flanges and are installed in a straight run of smooth pipe to avoid disturbance of flow patterns from fittings and valves.

Three kinds of orifice plates are used: concentric, eccentric, and segmental (Figure 5–8). The concentric orifice plate is the most common of the three types. As shown, the orifice is equidistant (concentric) to the inside diameter of the pipe. Flow through a sharp-edged orifice plate is characterized by a change in velocity. As the fluid passes through the orifice, the fluid converges, and the velocity of the fluid increases to a maximum value. At this point, the pressure is at a minimum value. As the fluid diverges to fill the entire pipe area, the velocity decreases back to the original value. The pressure increases to about 60% to 80% of the original input value. The pressure loss is irrecoverable; therefore, the output pressure will always be less than the input pressure. The pressures on both sides of the orifice are measured, resulting in a differential pressure that is proportional to the flow rate.

Segmental and eccentric orifice plates are functionally identical to the concentric orifice. The circular section of the segmental orifice is concentric with the pipe. The segmental portion of the orifice eliminates damming of foreign materials on the upstream side of the orifice when mounted in a horizontal pipe. Depending on the type of fluid being measured, the segmental section is placed on either the top or bottom of the horizontal pipe to increase the accuracy of the measurement. Eccentric orifice plates shift the edge of the orifice to the inside of the pipe wall. This design also prevents upstream damming and is used in the same way as the segmental orifice plate. Orifice plates have two distinct disadvantages: they cause a high, permanent pressure drop (outlet pressure will be 60% to 80% of inlet pressure) and they are subject to erosion, which will eventually cause inaccuracies in the measured differential pressure.

The venturi tube, illustrated in Figure 5–12, is the most accurate flow-sensing element when properly calibrated. The venturi tube has a converging conical inlet, a cylindrical throat, and a diverging recovery cone. It has no projections into the fluid, no sharp corners, and no sudden changes in contour. The inlet section decreases the area of the fluid stream, causing the velocity to increase and the pressure to decrease. The low pressure is measured in the center of the cylindrical throat where the pressure is at its lowest value, and neither the pressure nor the velocity changes. The recovery cone allows for the recovery of pressure such that total pressure loss is only 10% to 25%. The high pressure is measured upstream of the entrance cone. The major disadvantages of this type of flow detection are the high initial costs for installation and difficulty in installation and inspection.

FIGURE 5–12 Venturi tube.

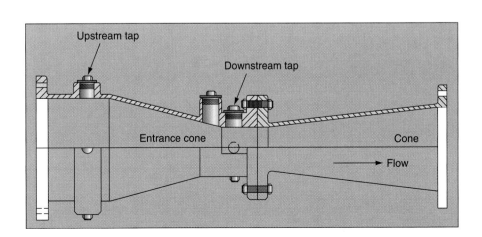

■ VELOCITY METERS

These meters operate with linearity with respect to the flow volume. There is no square-root relationship and their rangeability is greater. Velocity meters have a minimum sensitivity to viscosity changes when applied to a flow with a Reynold's number calculation greater than 10,000. Most velocity meters come with flanges, making them suitable piping arrangements to allow installation directly into pipelines.

5.8 Dall Flow Tube

The dall flow tube, illustrated in Figure 5–13, has a higher ratio of pressure developed to pressure lost than seen in the venturi flow tube. It is more compact and is commonly used in large-flow applications. The tube consists of a short, straight inlet section followed by an abrupt decrease in the inside diameter of the tube. This section, called the inlet shoulder, is followed by the converging inlet cone and a diverging exit cone. The two cones are separated by a slot or gap between the two cones. The low pressure is measured at the slotted throat (area between the two cones). The high pressure is measured at the upstream edge of the inlet shoulder. The dall flow tube is available in medium to very large sizes. The cost of the large sizes is normally less than that of a venturi flow tube. This type of flow tube has a pressure loss of about 5%.

5.9 Pitot Tube

The pitot tube, illustrated in Figure 5–14, is another primary flow element used to produce a differential pressure for flow detection. In its simplest form, it consists of a tube with an opening at the end. The small hole in the end is positioned so that it faces the flowing fluid. The velocity of the fluid at the opening of the tube decreases to zero. This provides for the high-pressure input to a differential pressure detector. A pressure tap provides the low-pressure input.

The pitot tube actually measures fluid velocity instead of fluid flow rate. However, volumetric flow rate can be obtained using the following equation: $Q = kAV$ (the same as before), where:

Q = volumetric flow rate

k = flow coefficient (approximately 0.8)

A = area of the flow cross-section

V = velocity of the flowing fluid

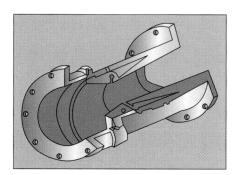

FIGURE 5–13   Dall flow tube.

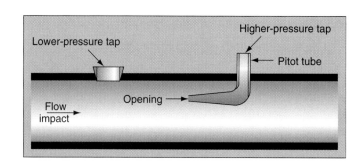

FIGURE 5–14   Pitot tube.

Head flow meters operate on the principle of placing a restriction in the line to cause a pressure drop. The differential pressure that is caused by the head is measured and converted to a flow measurement. The basic construction of various types of head flow detectors is summarized next.

Head Flow Meter Construction Summary

Orifice Plates

- Flat plates, 1/16- to 1/4-inch thick;
- Mounted between a pair of flanges; and
- Installed in a straight run of smooth pipe to avoid disturbance of flow patterns due to fittings and valves.

Venturi Tube

- A converging, conical inlet; a cylindrical throat; and a diverging recovery cone; and
- No projections into the fluid, no sharp corners, and no sudden changes in contour.

Dall Flow Tube

- A short, straight inlet section followed by an abrupt decrease in the inside diameter of the tube;
- An inlet shoulder followed by the converging inlet cone and a diverging exit cone; and
- Two cones separated by a slot or gap between the two cones.

Pitot Tube

- A tube with an opening at the end; and
- A small hole in the end positioned so that it faces the flowing fluid.

■ AREA FLOW METERS

The head causing the flow through an area meter is relatively constant such that the rate of flow is directly proportional to the metering area. The variation in area is produced by the rise and fall of a floating element. This type of flow meter must be mounted so that the floating element moves vertically and friction is minimal.

5.10 Rotameter

The rotameter, illustrated in Figure 5–15, is an area flow meter so named because the indicating element is a rotating float. The rotameter consists of a metal float and a conical glass tube, constructed such that the diameter increases with height. When there is no fluid passing through the rotameter, the float rests at the bottom of the tube. As fluid enters the tube, the higher density of the float causes the float to remain on the bottom. The space between the float and the tube allows for flow past the float. As flow increases in the tube, the pressure drop increases. When the pressure drop is sufficient, the float will rise to indicate the

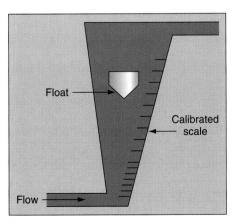

FIGURE 5–15 Rotameter using volumetric flow rate.

amount of flow. The higher the flow rate, the greater the pressure drop. The higher the pressure drop, the farther up the tube the float rises. The float should stay at a constant position at a constant flow rate. With a smooth float, fluctuations appear even when flow is constant. A float with slanted slots cut in the head maintains a constant position with respect to flow rate. This type of flow meter is usually used to measure low flow rates.

5.11 CORIOLIS METER

The Coriolis meter uses an obstructionless, U-shaped tube as a **sensor** and applies Newton's Second Law of Motion to determine flow rate. Inside the sensor housing, the sensor tube vibrates at its natural frequency (Figure 5–16). The sensor tube is driven by an electromagnetic drive coil located at the center of the bend in the tube that vibrates like a tuning fork.

The fluid flows into the sensor tube and is forced to take on the vertical momentum of the vibrating tube. When the tube is moving upward during half of its vibration cycle (Figure 5–17), the fluid flowing into the sensor resists being forced upward by pushing down on the tube.

The fluid flowing out of the sensor has an upward momentum from the motion of the tube. As it travels around the tube bend, the fluid resists changes in its vertical motion by pushing up on the tube, as shown in Figure 5–17. The difference in forces causes the sensor tube to twist (Figure 5–18). When the tube is moving downward during the second half of its vibration cycle, it twists in the opposite direction. This twisting characteristic is called the Coriolis effect.

Due to Newton's Second Law of Motion, the amount of sensor tube twist is directly proportional to the mass flow rate of the fluid flowing through the tube. Electromagnetic velocity detectors located on each side of the flow tube measure the velocity of the vibrating tube. **Mass flow rate** is determined by measuring the time difference exhibited by the velocity detector signals. During zero flow conditions, no tube twist occurs, resulting in no time difference between the two velocity signals. With flow, a twist occurs with a resulting time difference between the two velocity signals. This time difference is directly proportional to mass flow.

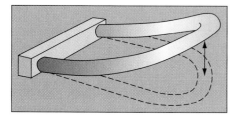

FIGURE 5–16 Vibrating Coriolis sensor tube.

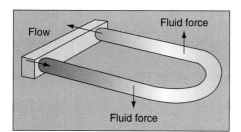

FIGURE 5–17 Fluid forces in a Coriolis sensor tube.

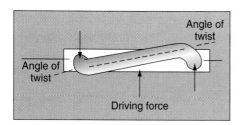

FIGURE 5–18 Coriolis effect.

5.12 HOT-WIRE ANEMOMETER

The hot-wire **anemometer,** principally used in gas flow measurement, consists of an electrically heated, fine platinum wire that is immersed into the flow. As the fluid velocity increases, the rate of heat flow from the heated wire to the flow stream increases. Thus, a cooling effect on the wire electrode occurs, causing its electrical resistance to change. In a constant-current anemometer, the fluid velocity is determined from a measurement of the resulting change in wire resistance. In a constant-resistance anemometer, fluid velocity is determined from the current needed to maintain a constant wire temperature and, thus, the resistance constant.

5.13 Electromagnetic Flow Meter

The electromagnetic flow meter is similar in principle to the generator. The rotor of the generator is replaced by a pipe placed between the poles of a magnet so that the flow of the fluid in the pipe is normal to the magnetic field. As the fluid flows through this magnetic field, an electromotive force is induced in it that will be mutually normal (perpendicular) to both the magnetic field and the motion of the fluid. This electromotive force can be measured with the aid of electrodes attached to the pipe and connected to a galvanometer or an equivalent. For a given magnetic field, the induced voltage will be proportional to the average velocity of the fluid. However, the fluid should have some degree of electrical conductivity.

5.14 Ultrasonic Flow Equipment

Devices such as ultrasonic flow equipment use the Doppler frequency shift of ultrasonic signals reflected from discontinuities in the fluid stream to obtain flow measurements. These discontinuities can be suspended solids, bubbles, or interfaces generated by turbulent eddies in the flow stream. The sensor is mounted on the outside of the pipe, and an ultrasonic beam from a piezoelectric crystal is transmitted through the pipe wall into the fluid at an angle to the flow stream. Signals reflected off flow disturbances are detected by a second piezoelectric crystal located in the same sensor. Transmitted and reflected signals are compared in an electrical circuit, and the corresponding frequency shift is proportional to the flow velocity.

Area Flow Meter Summary

Rotameter

- A metal float and a conical glass tube;
- Tube diameter increases with height;
- High-density float will remain on the bottom of tube with no flow;
- Space between the float and the tube allows for flow past the float; and
- As flow increases, the pressure drop increases; when the pressure drop is sufficient, the float rises to indicate the amount of flow.

Nutating Disc

- Circular disc attached to a central ball;
- A shaft is fastened to the ball and held in an inclined position by a cam or roller;
- Fluid enters an opening in the spherical wall on one side of the partition and leaves through the other side; and
- As the fluid flows through the chamber, the disc wobbles, or executes a nutating motion.

Coriolis Meter

- U-shaped tube;
- Mass flow measurement; and
- Proportional flow measurement.

Hot-Wire Anemometer

- Electrically heated, fine platinum wire immersed in flow;
- Wire is cooled as flow is increased; and
- Measures either change in wire resistance or heating current to determine flow.

Electromagnetic Flow Meter

- Magnetic field established around system pipe;
- Electromotive force induced in fluid as it flows through magnetic field; and
- Electromotive force measured with electrodes and is proportional to flow rate.

Ultrasonic Flow Equipment

- Uses Doppler frequency shift of ultrasonic signals reflected off discontinuities in fluid.

5.15 Mass Flow Meters

The demand for accurate flow measurements in mass-related processes (chemical, refining, heat transfer, etc.) has resulted in the design of mass flow meters. Various designs are available, but the one most widely used is the Coriolis meter. Its design is based upon the natural effect called the **Coriolis force.**

Coriolis meters are true mass meters that measure mass rate of flow directly as opposed to volumetric flow. Since the mass of flow does not change, the output is linear to the flow without having to be adjusted for variations in liquid properties, changes in temperature, and changes in pressure. Coriolis meters are available with various designs, but the one most used consists of a flow tube in a sensor housing, installed directly into the process, and connected to an electronics unit that may be located up to 500 feet away from the sensor.

Open-Channel Meters

The term *open channel* refers to any fluid routing that allows the fluid to flow with a free surface, such as in tunnels, nonpressured sewers, and rivers. Depth-related measurements are the most common type of measurement. This technique assumes that the instantaneous flow rate can be obtained by measuring the depth of the channel. Weirs and flumes are two of the oldest and most common open-channel measurement types.

Calibration

All flow meters require an initial calibration, but most of the time the instrument is calibrated by the manufacturer using the design specs provided. Field technicians must be able to perform their own calibrations to ensure accuracy. The need for continued calibration is determined by how well the meter fits its design criteria. Some liquids tend to be abrasive, corrosive, or erosive. Over a period of time, some parts of the sensing device may erode, thereby changing the original settings.

Some flow meters require special equipment for calibrating, but most suppliers will provide such equipment.

Maintenance

A number of factors can influence the need for maintenance of various devices. As stressed previously, matching the right flow meter with its correct application can reduce the need for maintenance. Flow meters with no moving parts will require less attention than those with moving parts, but all flow meters will eventually require some sort of maintenance.

Primary elements in differential flow meters can become plugged due to process content and may have to be cleaned or changed. Flow meters with moving parts require periodic inspection especially if the content being measured is dirty or viscous. Applications where coatings may occur also are a potential problem for motionless instruments such as the magnetic and ultrasonic meters. The electrodes of these meters can become insulated from the process and produce erroneous measurements. The probes should be cleaned and inspected periodically to ensure their proper operation.

5.16 Steam Flow Detection

Steam Flow Nozzle

Steam flow detection is normally accomplished through the use of a steam flow nozzle. There are occasions when we must be able to describe density compensation of a steam flow instrument and include the reason density compensation is required and identify the parameters used. The flow nozzle is commonly used for the measurement of steam flow and other high-velocity fluid flow measurements where erosion may occur. It is capable of measuring approximately 60% higher flow rates than an orifice plate with the same diameter. This is due to the streamlined contour of the throat, which is a distinct advantage for the measurement of high-velocity fluids. The flow nozzle requires less straight-run piping than does an orifice plate. However, the pressure drop is about the same for both. A typical flow nozzle is shown in Figure 5–19.

FIGURE 5–19   Steam flow nozzle.

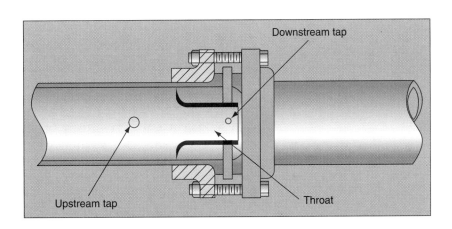

Since steam is considered to be a gas, changes in pressure and temperature greatly affect its density. The following equation gives the fundamental relationship for volumetric flow and mass flow:

$$\dot{V} = K\sqrt{\frac{\text{head loss}}{\rho}}h$$

and

$$\dot{m} = \dot{V}\rho$$

where

$\dot{V}$ = volumetric flow
K = constant relating to the ratio of pipe to orifice
h = differential pressure
ρ = density
$\dot{m}$ = mass flow

It is possible to substitute for density in the relationship using the following equation:

$$\rho = \frac{pm}{R\theta}$$

where

ρ = density
p = upstream pressure
m = molecular weight of the gas
θ = absolute temperature
R = gas constant

By substituting for density, the values are used by the electronic circuit to calculate the density automatically. Since steam temperature is relatively constant in most steam systems, upstream pressure is the only variable in the previous equation that changes as the system operates. If the other variables are hardwired, measuring the system pressure is all that is required for the electronics to calculate the fluid's density. As the previous equations demonstrate, temperature and pressure values can be used to electronically compensate flow for changes in density. A simple mass flow detection system is illustrated in Figure 5–20, where measurements of temperature and pressure are made with commonly used instruments.

For the precise measurement of gas flow (steam) at varying pressures and temperatures, it is necessary to determine the density, which is pressure and temperature dependent. From this value, we calculate the actual flow. The use of a computer is essential to measure flow with changing pressure or temperature. Figure 5–21 illustrates an example of a computer specifically designed for the measurement of gas flow. The computer is designed to accept input signals from commonly used

FIGURE 5-20 Simple mass flow detection system.

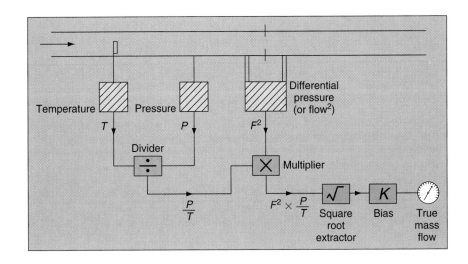

FIGURE 5-21 Gas flow computer system.

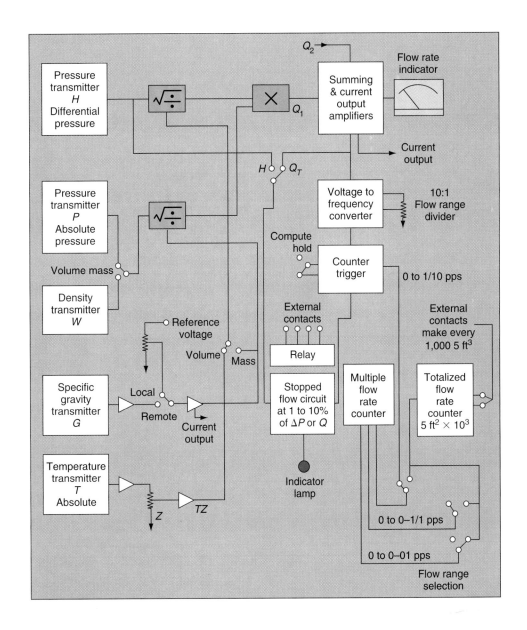

differential pressure detectors, or from density or pressure plus temperature sensors, and to provide an output that is proportional to the actual rate of flow. The computer has an accuracy better than +0.1% at flow rates of 10% to 100%.

■ ENVIRONMENTAL CONCERNS

As previously discussed, the density of the fluid whose flow is to be measured can have a large effect on flow-sensing instrumentation. The effect of density is most important when the flow-sensing instrumentation is measuring gas flows, such as steam. Since the density of a gas is directly affected by temperature and pressure, any changes in either of these parameters will have a direct effect on the measured flow. Therefore, any changes in fluid temperature or pressure must be compensated for to achieve an accurate measurement of flow. Ambient temperature variations will affect the accuracy and reliability of flow-sensing instrumentation. Variations in ambient temperature can directly affect the resistance of components in the instrumentation circuitry and, therefore, affect the calibration of electric/electronic equipment. The effects of temperature variations are reduced by the design of the circuitry and by maintaining the flow-sensing instrumentation in the proper environment. The presence of humidity will also affect most electrical equipment, especially electronic equipment. High humidity causes moisture to collect on the equipment, causing short circuits, grounds, and corrosion, which, in turn, may damage components. The effects due to humidity are controlled by maintaining the equipment in the proper environment. The density of the fluid, ambient temperature, and humidity are the three factors that can affect the accuracy and reliability of flow-sensing instrumentation as shown in Figure 5–22 for steam measurement.

■ EQUIVALENT MEASUREMENT VARIABLES

Often the measurement devices used do not reflect the actual flow rates that operations may use to observe and control the process. It is helpful if we have an understanding of conversion tables and know how to utilize them to obtain the desired measured variable. For example, a differential pressure measuring device has its differential pressure measured in H_2O. The following process shows how a conversion to GPH proves its equivalent measuring units:

$$Q = V\left(\frac{in}{sec}\right) \times A(in^2) \quad \text{(Bernoulli's principle)}$$

$$Q = (\sqrt{H})\frac{in}{sec} \times A(in^2) \times k_{(constant)(bias)} \quad \text{(Bernoulli's principle applied)}$$

$$Q = flow\frac{in^3}{sec}$$

FIGURE 5–22 Differential steam venturi.

5.17 Volumetric Conversions

The previous equations demonstrate how a volumetric flow rate is obtained from a differential pressure measurement. Once a volumetric flow rate is obtained, conversion to equivalent measuring parameters is a function of applied math. For this example, we will arbitrarily select some in^3/sec value and derive a GPH process variable from it:

$$2{,}748\frac{\text{in}^3}{\text{sec}} \;\rightarrow$$

$$2{,}748\frac{\text{in}^3}{\text{sec}} \times 60\frac{\text{sec}}{\text{min}} = 164{,}880\frac{\text{in}^3}{\text{min}}$$

$$164{,}880\frac{\text{in}^3}{\text{min}} \times 60\frac{\text{min}}{\text{hour}} = 9{,}892{,}800\frac{\text{in}^3}{\text{hour}}$$

$$\frac{9{,}892{,}800\dfrac{\text{in}^3}{\text{hour}}}{1{,}728\dfrac{\text{in}^3}{\text{ft}^3}} = 5{,}725\frac{\text{ft}^3}{\text{hour}}$$

$$5{,}725\frac{\text{ft}^3}{\text{hour}} \times 7.4805\frac{\text{gal}}{\text{ft}^3} = 42{,}825.86\frac{\text{gal}}{\text{hour}}\;(\text{GPH})$$

5.18 Mass Flow Conversions

There are occasions when we must be able to convert from some mass flow rate (lb/hour) to its equivalent GPH variable as well. The following example will utilize a process reading of 3,240 (lb/hour) and convert it to GPH:

$$\frac{62.43\frac{\text{lb}}{\text{ft}^3}}{7.4805\frac{\text{gal}}{\text{ft}^3}} = 8.35\frac{\text{lb}}{\text{gal}}$$

$$\frac{3,240\frac{\text{lb}}{\text{hour}}}{8.35\frac{\text{lb}}{\text{gal}}} = 388\frac{\text{gal}}{\text{hour}}\ (\text{GPH})$$

FIGURE 5–23 Micro Motion elite sensor with remote-mount transmitter for gas flow, installed up.

FIGURE 5–24 Dwyer Visa-Float flow meters used for indication and control of nitrogen purge to electrical panels.

FIGURE 5-25 Vortex flow meter; shredder inside flow tube measures frequency.

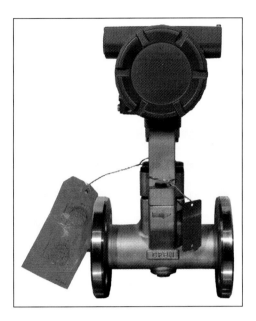

SUMMARY

Flow meters rely on many different physical and mathematical properties to accurately measure flow. The measured properties can be altered by introducing errors into the measurement loop through installation and/or calibration. This chapter provided an overview of the physics and installation requirements involved so that any flow-measuring device can be installed and calibrated correctly.

REVIEW QUESTIONS

1. Explain Bernoulli's principle mathematically. Then demonstrate how a volumetric flow rate can be determined from a velocity measurement.

2. When velocity increases in a flow line, will the corresponding pressure increase or decrease?

3. Velocity, V, as a measured variable utilizing a differential pressure measurement device, is proportional and equal to differential pressure by what equation?

4. Is laminar flow or turbulent flow desired for differential pressure measurement?

5. What type of orifice plate is commonly used to measure slurries?

6. What is the point downstream of an orifice plate where fluid velocity is greatest?

7. What would be the output, in mA, of a differential pressure transmitter configured for a square-rooted output with 60% (of span) applied pressure?

8. What type of instrument is the most accurate flow-sensing element when properly calibrated?

9. Do velocity meters operate linearly or nonlinearly with respect to flow volume?

10. How is an area flow meter usually identified visually?

11. The vibration that is measured and caused due to a flow rate affecting a U-tube is known as what type of principle?

chapter 6

Fundamentals of Liquid Level

■ OUTLINE

■ OVERVIEW

A wide variety of devices are used to measure and record level. The applications of devices may differ, but the operating principles—the fundamentals—are explained in this chapter. Calibration measurement errors can be introduced through improper or uncompensated mounting procedures. If the installation of the device is not performed correctly, the installer must be aware of compensations that may be needed. This chapter provides an excellent overview and details of the process characteristics that must be understood to accurately measure and record level.

■ OBJECTIVES

After completing this chapter, you should be able to:

- Describe level measurement properties and measurement principles.
- Show mathematically level relationships.
- Describe how differential pressure can be used to derive a level measurement.
- List various devices that can be used to measure level.
- Describe differential pressure applications to measure level.
- Explain the relationships and use of specific gravity.
- Explain Archimedes' law.
- Describe the operation of capacitance, resistance, radar, and ultrasonic measurement.
- Explain why compensation may be needed for steam measurement.

■ INTRODUCTION

6.1 Level Fundamentals

Liquid level measuring devices are classified into two groups: direct method and inferred method. An example of the direct method is the dipstick in your car, which measures the height of the oil in the oil pan. An example of the inferred method is a pressure gauge at the bottom of a tank that measures the hydrostatic head pressure from the height of the liquid.

Liquid level measurements may consist of several methods that either infer or directly measure the process. This chapter discusses the following methods: gauge glass, ball float, chain float, magnetic bond, conductivity probe, differential pressure (ΔP), capacitive, radar, and ultrasonic level.

■ LEVEL MEASUREMENT APPLICATIONS

6.2 Gauge Glass

The gauge glass method is a very simple means by which liquid level is measured in a vessel (Figure 6–1). In the gauge glass method, a transparent tube is attached to the bottom and top (the top connection is not needed in a tank open to atmosphere) of the tank. The height of the liquid in the tube is equal to the height of water in the tank.

Figure 6–1(a) shows a gauge glass that is used for vessels with liquid at ambient temperature and pressure conditions. Figure 6–1(b) shows a gauge glass that is used for vessels

FIGURE 6–1 Transparent tube
level measurement.

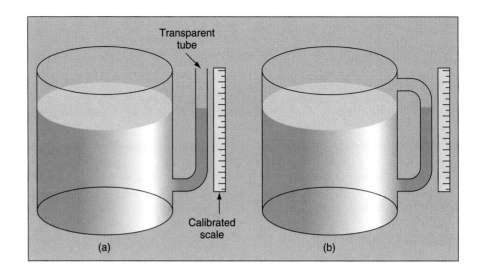

with liquid at an elevated pressure or a partial vacuum. Notice that the gauge glasses in Figure 6–1 effectively form a U-tube manometer in which the liquid seeks its own level due to the pressure of the liquid in the vessel.

Gauge glasses made from tubular glass or plastic are used for service up to 450 PSIG and 400°F (Figure 6–2). A different type of gauge glass is used to measure the level of a vessel at higher temperatures and pressures. The type of gauge glass utilized in this instance has a body made of metal with a heavy glass or quartz section for visual observation of the liquid level. The glass section is usually flat to provide strength and safety. Figure 6–2 illustrates a typical transparent gauge glass.

In another type of gauge glass, the reflex gauge glass (Figure 6–3), one side of the glass section is prism-shaped. The glass is molded such that one side has 90° angles that run lengthwise. Light rays strike the outer surface of the glass at a 90° angle. The light rays travel through the glass, striking the inner side of the glass at a 45° angle. The presence or absence of liquid in the chamber determines if the light rays are refracted into the chamber or reflected back to the outer surface of the glass.

FIGURE 6–2 Gauge glass for level measurement.

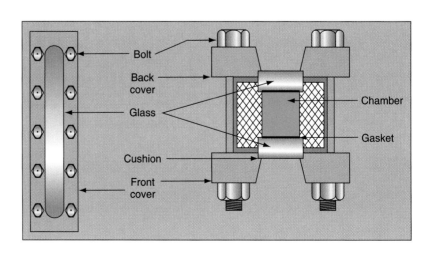

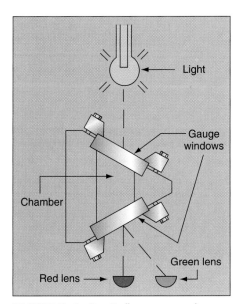

FIGURE 6-3 Reflex gauge glass.

When the liquid is at an intermediate level in the gauge glass, the light rays encounter an air-glass interface in one portion of the chamber and a water-glass interface in the other portion of the chamber. Where an air-glass interface exists, the light rays are reflected back to the outer surface of the glass because the critical angle for light to pass from air to glass is 42°. This causes the gauge glass to appear silvery-white. In the portion of the chamber with the water-glass interface, the light is refracted into the chamber by the prisms. Reflection of the light back to the outer surface of the gauge glass does not occur because the critical angle for light to pass from glass to water is 62°. This results in the glass appearing black, since it is possible to see through the water to the walls of the chamber, which are painted black.

A third type of gauge glass is the refraction type (Figure 6–4). This type is especially useful in areas of reduced lighting; lights are usually attached to the gauge glass. Operation is based on the principle that the bending of light, or *refraction,* will be different as light passes through various media. Light is bent, or refracted, to a greater extent in water than in steam. For the portion of the chamber that contains steam, the light rays travel relatively straight, and the red lens is illuminated. For the portion of the chamber that contains water, the light rays are bent, causing the green lens to be illuminated. The portion of the gauge containing water appears green; the portion of the gauge from that level upward appears red.

6.3 Ball Float

The ball float method is a direct-reading, liquid level mechanism. The most practical design for the float is a hollow metal ball or sphere. However, there are no restrictions to the size, shape, or material used. The design consists of a ball float attached to a rod, which, in turn, is connected to a rotating shaft that indicates level on a calibrated scale (Figure 6–5).

The operation of the ball float is simple: The ball floats on top of the liquid in the tank. If the liquid level changes, the float will follow and change the position of the pointer attached to the rotating shaft. The travel of the ball float is limited by its design to be within ±30° from

FIGURE 6-4 Overhead view, refraction gauge glass.

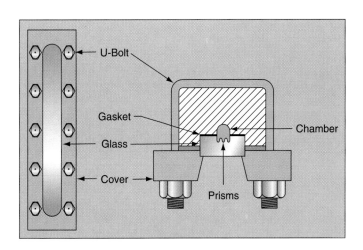

FIGURE 6-5 Ball float level.

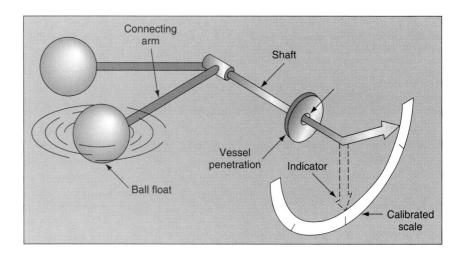

the horizontal plane, resulting in optimum response and performance. The actual level range is determined by the length of the connecting arm. The stuffing box is incorporated to form a watertight seal around the shaft to prevent leakage from the vessel.

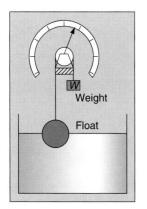

FIGURE 6-6 Chain float gauge.

6.4 Chain Float

This type of float gauge has a float ranging in size up to 12 inches in diameter and is used where small-level limitations imposed by ball floats must be exceeded. The range of level measured will be limited only by the size of the vessel. The operation of the chain float is similiar to that of the ball float except in the method of positioning the pointer and in its connection to the position indication. The float is connected to a rotating element by a chain with a weight attached to the other end to provide a means of keeping the chain taut during changes in level.

6.5 Magnetic Bond Method

The magnetic bond method was developed to overcome the problems of cages and stuffing boxes (see Figure 6–6). The magnetic bond mechanism consists of a magnetic float that rises and falls with changes in level. The float travels outside of a nonmagnetic tube that houses an inner magnet connected to a level indicator. When the float rises and falls, the outer magnet attracts the inner magnet, causing the inner magnet to follow the level within the vessel (Figure 6–7).

6.6 Conductivity Probe Method

Figure 6–8 illustrates a conductivity probe level detection system. It consists of one or more level detectors, an operating relay, and a controller. When the liquid makes contact with any of the electrodes, an electric current will flow between the electrode and ground. The current energizes a relay, which causes the relay contacts to open or close depending on the state of the process involved. The relay, in turn, will actuate an alarm, a pump, a control valve, or all three. A typical system has three probes: a low-level probe, a high-level probe, and a high-level alarm probe.

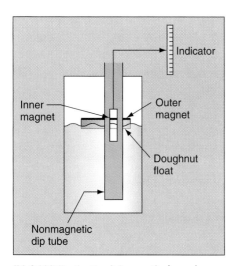

FIGURE 6-7 Magnetic bond detector.

FIGURE 6–8 Conductivity level detection.

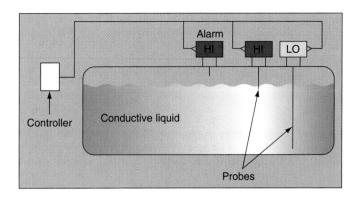

Today's typical process plant contains various tanks, vessels, and reservoirs, all of which will store, at some time, various stages of the process. It is essential to record accurate measurements of the vessels to maintain an automatic working process. The majority of the tanks will contain a liquid of some form, but some may be used to contain solids. Level measurement is an integral part of process control and may be used in a wide variety of applications.

One of the most common ways to measure a liquid level is to use a pressure gauge to measure the pressure of a liquid, and knowing the specific gravity of that liquid, to obtain the proper level. This method uses the differential pressure measurement device to detect the desired level.

■ DIFFERENTIAL PRESSURE LEVEL DETECTORS

The differential pressure (ΔP) detector method of liquid level measurement uses a ΔP detector connected to the bottom of the tank being monitored. The higher pressure, caused by the fluid in the tank, is compared to a lower reference pressure (usually atmospheric). This comparison takes place in the ΔP detector. Figure 6–9 illustrates a typical differential pressure detector attached to an open tank.

6.7 Open-Tank Measurement

The tank is open to the atmosphere; therefore, it is necessary to use only the high-pressure (HP) connection on the ΔP transmitter. The low-pressure (LP) side is vented to the atmosphere; therefore, the pressure differential is the hydrostatic head, or weight, of the liquid in the tank. The maximum level that can be measured by the ΔP transmitter is determined by the maximum height of liquid above the transmitter. The minimum level that can be measured is determined by the point where the transmitter is connected to the tank (Figure 6–9).

6.8 Closed-Tank, Dry Reference Leg Measurement

Not all tanks and vessels are open to the atmosphere. Many are totally enclosed to prevent vapors or steam from escaping, or to allow

FIGURE 6–9 Open-tank differential pressure measurement.

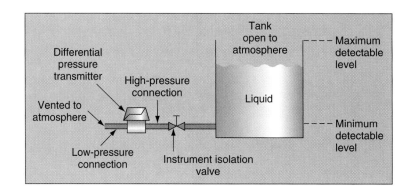

pressurizing the contents of the tank. When measuring the level in a tank that is pressurized, or the level that can become pressurized by vapor pressure from the liquid, both the high-pressure and low-pressure sides of the ΔP transmitter must be connected (Figure 6–10).

The high-pressure connection is connected to the tank at or below the lower range value to be measured. The low-pressure side is connected to a "reference leg" that is connected at or above the upper range value to be measured. The reference leg is pressurized by the gas or vapor pressure, but no liquid is permitted to remain in the reference leg. The reference leg must be maintained dry so that there is no liquid head pressure on the low-pressure side of the transmitter. The high-pressure side is exposed to the hydrostatic head of the liquid plus the gas or vapor pressure exerted on the liquid's surface. The gas or vapor pressure is equally applied to the low- and high-pressure sides. Therefore, the output of the ΔP transmitter is directly proportional to the hydrostatic head pressure—that is, the level in the tank.

6.9 Closed-Tank, Filled (Wet) Reference Leg Measurement

A closed-tank measurement that has a filled reference leg is connected in the same manner as a dry reference leg, but the measurement values read negative (inversely proportional). This is due to an excess pressure on the low-reference leg, which drives the measured variable into the negative range. The filled leg maintains an additional, or unwanted, pressure on the "low" side of the differential pressure sensor, which causes the signal to read increasing pressure, but is maintained in the negative range.

FIGURE 6–10 Closed-tank, dry reference leg measurement.

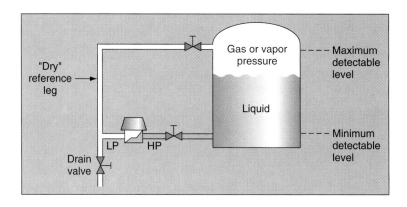

6.10 Inverse Liquid Level Measurement

There are occasions when the differential pressure-measuring device can be used to measure the differential pressure directly, but the measurement arrangement of the device is installed to provide an inverse measurement of level.

One such occasion is when the measuring device installed is not capable of measuring "negative" values of differential pressure. By installing the device to measure increasing amounts of pressure, the pressure taps are reversed. The high-pressure tap of the process tubing is connected to the low-pressure port of the differential pressure device. Likewise, the low-pressure tap of the process connection is attached to the high-pressure port of the differential pressure-measuring device (Figure 6–11). Also, if the tank contains a condensible fluid, such as steam, a slightly different arrangement can be used.

In applications with condensible fluids, condensation is greatly increased in the reference leg. To compensate for this effect, the reference leg is filled with the same fluid as in the tank. The liquid in the reference leg applies a hydrostatic head to the high-pressure side of the transmitter, and the value of this level is constant as long as the reference leg is maintained full. If this pressure remains constant, any change in ΔP is due to a change on the low-pressure side of the transmitter (Figure 6–11).

The filled reference leg applies a hydrostatic pressure to the high-pressure side of the transmitter, which is equal to the maximum level to be measured. The ΔP transmitter is exposed to equal pressure on the high- and low-pressure sides when the liquid level is at its maximum; therefore, the differential pressure is zero. As the tank level goes down, the pressure applied to the low-pressure side goes down also, and the differential pressure increases. As a result, the differential pressure and the transmitter output are inversely proportional to the tank level.

FIGURE 6–11   Closed-tank, filled reference leg, inverse measurement.

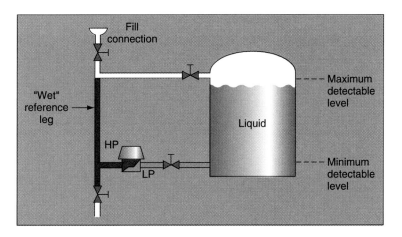

6.11 Continuous and Point Level Measurement

Level-sensing devices can be grouped into two categories: point level measurement and continuous level measurement. Point level (level switch) measurement monitors a specific level height and sends a discrete signal when this point is reached. Discrete signals are used for

Lesson 8-12

alarm conditions, motor starting/stopping, shutdown signals, and several more, but the trend is for continuous level monitors to have control over all process systems.

A continuous level sensor monitors the level height of a liquid over a wider range rather than at a single point as in discrete point control. The continuous level monitor (transmitter and sensor) provide a working range that controllers can "see"; with discrete point (level switch) control, controllers are working in the blind until the discrete point is obtained. Circumstances for each control loop will determine the type of control loop that is required, but field technicians must be the interface between controllers and control loops.

Current trends point toward a level measurement device used in industrial applications that is relatively inexpensive, easy to install, and easy to calibrate and maintain. The differential pressure transmitter that records head pressure is one such device. Head pressure uses the pressure of a liquid, developed by the liquid's density and height above the measuring point, to determine a pressure value equivalent to inches of water (head pressure). From the pressure lesson covered previously, we know that 27.7 inches of water is equal to 1 PSI:

$$27.7'' \ H_2O = 1 \ PSI$$

6.12 Pressure and Head Calculations

Since all liquids can be calculated to express their weight in depth (height) and density, we can use the mathematical relationship $P = S \times H$ to express level, where P is the pressure of a liquid, S is the specific gravity of the liquid, and H is the height of the liquid. The pleasant result is that P can be expressed in inches of water column ($''H_2O$) or in PSI as needed.

For example, use Figure 6–12 to calculate the equivalent pressure in PSI and $''H_2O$. The tank is filled with gasoline, which has a specific gravity of 0.67. To obtain the pressure of a liquid, use the equation $P = W \times D$ (P = pressure (PSI), W = weight density, D = depth). To calculate the weight density, recall that the density of water = 62.43 lb/cubic foot.

$$P_{\text{(pressure)}} = W_{\text{(weight density)}} \times D_{\text{(depth)}}$$

$$P = 62.43 \frac{\text{lb}}{\text{ft}^3} \times 0.67_{\text{(specific gravity)}} \times 15 \ \text{ft}$$

$$P = 627.42 \frac{\text{lb}}{\text{ft}^2}$$

Since pounds per square foot (lb/ft²) is too large for most pressure instruments, we must convert to pounds per square inch:

$$P = \frac{627.42 \frac{\text{lb}}{\text{ft}^2}}{144 \frac{\text{in}^2}{\text{ft}^2}}$$

$$P = 4.36 \frac{\text{lb}}{\text{in}^2} \text{(PSI)}$$

FIGURE 6–12   Head pressure measurement.

Most level instruments are calibrated in inches of water ($"H_2O$), so let's now calculate the calibrated pressure in inches of H_2O using the equation $P = S \times H$. Remember, the specific gravity is needed also.

$$P = S_{(\text{specific gravity})} \times H_{(\text{height})}$$

$$P = 0.67 \times 15 \text{ ft} \times 12\frac{\text{in}}{\text{ft}}$$

$$P = 0.67 \times 180 \text{ in}$$

$$P = 120.6 \text{ in}$$

Check your calculations by converting directly from PSI to $"H_2O$. Recall that one PSI of atmospheric pressure is equal to 27.7 inches of water:

$$P = 4.36 \text{ PSI} \times 27.7\frac{\text{in}}{\text{PSI}}$$

$$P = 20.77 \text{ in (difference is round off error)}$$

Most level instrument equipment will be calibrated in $"H_2O$; therefore, a more useful way to calculate the pressure of a liquid is to use the head pressure method ($P = S \times H$). This way of calibrating the pressure of a liquid actually simplifies the process by figuring the calibration range and pressure at the same time. This method can be remembered as the "tape measure rule," which implies what its name suggests. Simply measure the height of the level range needed in inches, multiply the height by the specific gravity, and the result is obtained in an inches measurement.

6.13 Instrument Elevation and Suppression Errors

The previous example uses a differential pressure (DP) sensor to sense level changes and the approach is fairly straightforward. Sometimes other considerations that will involve the installation and maintenance of differential pressure level sensors must be taken into consideration. The mounting location of DP sensors can actually play a part in the calibrated measurement of the level. If the DP sensor is located below the tank bottom, extra pressure will be on the DP cell simply from the pressure/weight of the liquid in the related tank/tubing to the DP sensor. This is an elevation problem, which means that the pressure of the liquid elevates the output. **Elevation** is defined as additional or unwanted pressure on the low side of a DP sensor. The term normally applies to a condition that causes the 4-mA output of a DP cell to be low when there is 0% level, so the output is adjusted to zero. In addition, sealing liquids, wet legs, filled leg applications, and other conditions may place unwanted pressure on the low-pressure side or the high-pressure side of the sensor and either elevate or suppress the output. The DP sensor must take this into consideration and must be elevated or suppressed to measure a true level. **Suppression** is defined as the lowering of the output caused by additional or unwanted pressure on the high side of a DP cell. The term is normally used for a condition that has the 4-mA output high when there is 0% level.

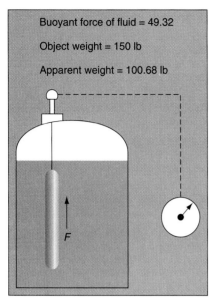

Buoyant force of fluid = 49.32

Object weight = 150 lb

Apparent weight = 100.68 lb

F

FIGURE 6–13 Archimedes's law.

ARCHIMEDES' LAW

6.14 Buoyant Force

Another common method of measuring level uses buoyancy (float switch). The basic float switch is a simple float that changes level with the level of the fluid. A recording and/or measuring device simply measures the change in level of the float to a reference point and transmits the output. Buoyant force is always directed in a way to force an object out of the liquid in which it is submerged. This commonsense principle is a form of Archimedes' law on buoyant force.

Archimedes' law states that when an object is placed into a fluid, the object will be subject to a buoyant force equal to the weight of the fluid displaced by the object. For example, let's study the displacement and buoyant force of a fluid on an object (1 ft³ volume) as shown in Figure 6–13.

Object weight = 150 lb

Alcohol-filled tank (specific gravity = 0.79)

Use the weight of an equivalent amount of water to find the buoyant force:

$$62.43 \, ^{lb}/_{ft^3} \times 0.79 = 49.32 \, ^{lb}/_{ft^3}$$

Applying Archimedes' law, we can find an apparent weight of the object by subtracting the buoyant force from the real weight when submerged:

$$150.00 \, ^{lb}/_{ft^3} - 49.32 \, ^{lb}/_{ft^3} = 100.68 \, ^{lb}/_{ft^3}$$

This apparent weight is only *apparent weight*—the object still weighs 150 pounds but the buoyant force of the liquid provides a weight that is apparent—not true weight. Using the buoyant force principle, we can measure a fluid's level. If the fluid level is below or at the bottom of the displacer tube (float), the displacer will register its full weight on the sensor: 150 pounds. As the fluid level rises, the displacer weight will decrease in exact proportions. As the level reaches 50% of the displacer length, the displacer will have an apparent weight determined by the following:

150 pounds per full volume implies that:

$$50\% \text{ of } 150 \text{ lb} = 0.50 \times 150 \text{ lb} = 75 \text{ lb}_{\text{(portion not submerged)}}$$

There is also the apparent weight of the submerged portion:

$$50\% \text{ of } 100.68 \text{ lb}_{\text{(apparent weight)}} = 0.50 \times 100.68 = 50.34 \text{ lb}$$

Therefore:

$$\text{Total} = 75 \text{ lb} + 50.34 \text{ lb} = 125.34 \text{ lb}$$

To further explain, the nonsubmerged portion is exactly half the weight of the entire displacer when it is not submerged. Likewise, the submerged portion is exactly half the apparent weight if it all was submerged. The total simply adds the two together. Likewise, as the level increases, the apparent weight will decrease proportionately.

It is important to realize that the level can only be detected along the length of the displacer. When the level is below or above the dis-

Test 3-25
Lesson 8-34

placer, the output of the transmitter will be maximum or minimum depending upon calibration.

As you may have guessed, a displacer can also detect the density of a liquid. Liquids with greater weight densities will cause more of an apparent weight change. If the tank was filled with water, the buoyant force would equal 62.43 pounds, the apparent weight of the displacer would be 150 lb − 62.43 lb =87.57 lb. (One related note: A displacer is often called a *float*. A float [displacer] never floats.)

■ CAPACITANCE MEASUREMENT

6.15 Capacitance Level Sensor

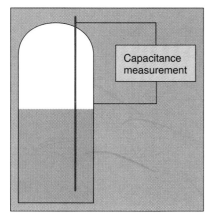

FIGURE 6–14 Capacitance measurement.

Another method for measuring level is to use a capacitance level sensor (Figure 6–14). A capacitance sensor probe is inserted into the tank and senses level change by related capacitive and resistive values of the liquid. The result is an accurate level measurement. Electrical capacitance is a fundamental concept for electricians that helps to explain the level measuring form of capacitance. Capacitance is the storing of an electrical charge by two metal plates separated by a **dielectric.** A metal rod (plate) is inserted into a tank and serves as one of the capacitance plates and the tank wall serves as another. When the tank is empty, the dielectric is air. When the tank is filled to some level, the dielectric changes due to the fluid becoming the dielectric and changing the original value. It is important to understand that the fluid can have a better or worse dielectric rating; the change is the function calculated by the sensor.

■ RESISTANCE MEASUREMENT

6.16 Resistance Probes

We can use fluids that have a poor dielectric rating to act as a conductor to detect level. As the fluid of a tank rises, it will reach the **probes** set in the tank and allow a sensor to record when the fluid level has reached the level of the probes. This method is good for monitoring the point at which the probes are located since the amount of liquid above or below the probe cannot be determined.

6.17 Radar and Ultrasonic Measurement

Other level measurement devices and their applications are unique to their respective measuring function. Radiation level measurement utilizing radar and ultrasonic applications to derive level measurement are unique applications and a brief mention of their respective functions is warranted here. Each application is based upon a material's tendency to absorb or reflect radiation. They have no moving components for calibration and sometimes require no physical contact with the process measured (Figure 6–15). The main advantage of a radiation-based level gauge is the absence of moving parts and the ability to detect level without making physical contact with the process fluid. Because they can, in effect, "see" through solid tank walls, nuclear radiation gauges are perhaps the ultimate in noncontact sensing devices.

FIGURE 6–15 "Time of flight" radar measurement.

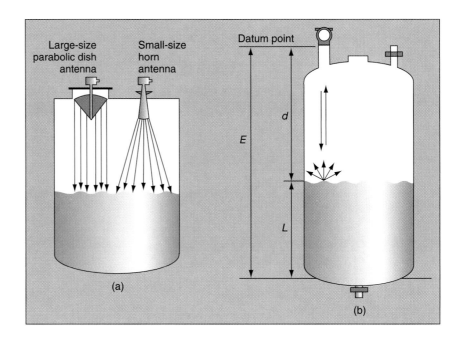

They are expensive and require a gamma radiation source, so they are often employed only as a last choice.

Radar measurement consists of a transmitter, an antenna, a receiver, and an operator interface. They may be a contact or no contact with the process and the application will be determined before installation. Radar sensors consist of a transmitter, an antenna, a receiver with a signal processor, and an operator interface. The transmitter is mounted on top of the vessel. Its solid-state oscillator sends out an electromagnetic wave (using a selected carrier frequency and waveform) aimed downward at the surface of the process fluid in the tank. The frequency used is generally 10 GHz. A portion of the transmitted wave is reflected back to the antenna, where it is collected and routed to the receiver. A microprocessor calculates the time of flight and calculates the resulting level from the time of flight measurement.

Time of flight is the period between the transmission of the radar pulse and the reception of the return echo and it is determined by the radar detector, which is simultaneously exposed to both the sent and the reflected signals. The detector output is based on the difference. The **frequency**-modulated (FM) signal varies from 0 to 200 Hz as the distance to the process fluid surface varies between 0 and 200 feet. Because this measurement takes place in the frequency domain, it is reasonably free of noise interference. Radar beams can penetrate plastic and fiberglass; therefore, noncontact radar gauges can be isolated from the process vapors by a seal. Contact radar gauges send a pulse down a wire to the vapor-liquid interface. There, a sudden change in the dielectric constant causes the signal to be partially reflected. The time-of-flight is then measured (Figure 6–16). The unreflected portion travels on to the end of the probe and provides a zero-level reference signal.

FIGURE 6–16 "Time of flight" applications.

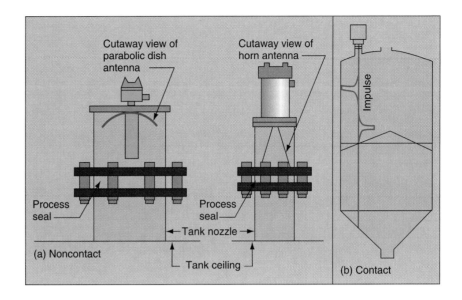

The frequency range of audible sound is 9 to 10 kHz, slightly below the 20- to 45-kHz range used by industrial-level gages. The velocity of an ultrasonic pulse varies with both the substance through which it travels and the temperature of that substance. This means that if the speed of sound is to be used in measuring a level (distance or position), the substance through which it travels must be well known and its temperature variations must be measured and compensated for. When mounted on the top of the tank, the sensor detects the depth of the vapor space. Accurate knowledge of the shape of the tank's cross section is required to determine the volume of liquid. When detecting the interface between two liquids, such as the hydrocarbon–brine interface in a salt dome storage well, the transducer is lowered down to the bottom of the well and the ultrasonic pulse is sent up through the heavy brine layer to the interface. The time it takes for the echo to return is an indication of the location of the interface (Figure 6–17). Most modern ultrasonic instruments include temperature compensation and filters for data processing and response time; some even provide self-calibration.

FIGURE 6–17 Ultrasonic level measurement.

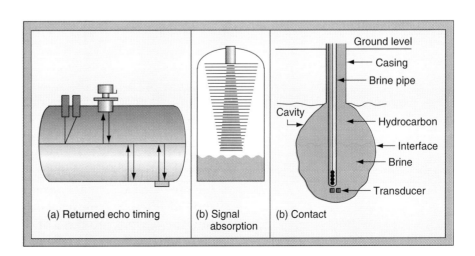

■ DENSITY COMPENSATION

Often the difference in temperature between the reference legs and the temperature of the process requires compensation for accurate level measurement. If a vapor with a significant density exists in the reference legs above the liquid measured, the hydrostatic pressure added needs to be considered if accurate transmitter output is required. To compensate for density differences, we need to understand why the compensation is needed and know how to accomplish this task. Before examining an example that shows the effects of density, we must define the unit *specific volume*. Specific volume is defined as volume per unit mass, as shown in the following equations.

6.18 Specific Volume

$$\text{Specific volume} = \frac{\text{Volume}}{\text{Mass}}$$

Specific volume is the reciprocal of density, as shown in the following equation:

$$\text{Specific volume} = \frac{1}{\text{Density}}$$

Specific volume is the standard unit used in working with vapors and steam that have low values of density. For the applications that involve water and steam, specific volume can be found using "Saturated Steam Tables," which list the specific volumes for water and saturated steam at different pressures and temperatures. The density of steam (or vapor) above the liquid level will have an effect on the weight of the steam or vapor bubble and the hydrostatic head pressure. As the density of the steam or vapor increases, the weight increases and causes an increase in hydrostatic head even though the actual level of the tank has not changed. The larger the steam bubble, the greater the change in hydrostatic head pressure. Figure 6–18 illustrates a vessel in which the water is at saturated boiling conditions.

A condensing pot at the top of the reference leg is incorporated to condense the steam and maintain the reference leg filled. As previously stated, the effect of the steam vapor pressure is cancelled at the

FIGURE 6–18 Effects of fluid density.

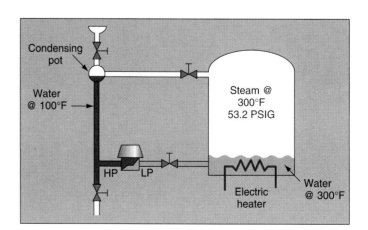

ΔP transmitter due to the fact that this pressure is applied equally to both the low- and high-pressure sides of the transmitter. The differential pressure to the transmitter is due only to hydrostatic head pressure, as stated in the following equation:

$$\text{Hydrostatic head pressure} = \text{Density} \times \text{Height}$$

6.19 Reference Leg Temperature Considerations

When the level to be measured is in a pressurized tank at elevated temperatures, a number of additional consequences must be considered. As the temperature of the fluid in the tank is increased, the density of the fluid decreases. As the fluid's density decreases, the fluid expands, occupying more volume. Even though the density is less, the mass of the fluid in the tank is the same. The problem encountered is that, as the fluid in the tank is heated and cooled, the density of the fluid changes but the reference leg density remains relatively constant, which causes the indicated level to remain constant. The density of the fluid in the reference leg is dependent upon the ambient temperature of the room in which the tank is located; therefore, it is relatively constant and independent of tank temperature. If the fluid in the tank changes temperature and, therefore, density, some means of density compensation must be incorporated to produce an accurate indication of tank level. This is the problem encountered when measuring pressurized water level or steam generator water level in pressurized water reactors, and when measuring reactor vessel water level in boiling water reactors.

6.20 Pressurizer Level Instruments

Figure 6–19 shows a typical pressurizer level system. Pressurizer temperature is held fairly constant during normal operation. The ΔP detector for level is calibrated with the pressurizer hot, and the effects of density changes do not occur. The pressurizer will not always be hot. It may be cooled down for nonoperating maintenance conditions, in which case a second ΔP detector, calibrated for level measurement at low temperatures, replaces the normal <ΔP detector. The density has not really been compensated for; it has actually been aligned out of the instrument by calibration. Density compensation can also be accomplished through electronic circuitry. Some systems compensate for density changes automatically through the design of the level detection circuitry. Other applications compensate for density by manually adjusting inputs to the circuit as the pressurizer cools down and depressurizes, or during heat-up and pressurization. Calibration charts are also available to correct indications for changes in reference leg temperature.

6.21 Steam Generator Level Instrument

Figure 6–20 illustrates a typical steam generator level detection arrangement. The ΔP detector measures actual differential pressure. A

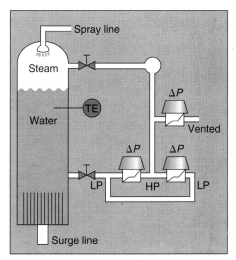

FIGURE 6–19 Pressurizer level measurement.

FIGURE 6–20 Steam generator level system.

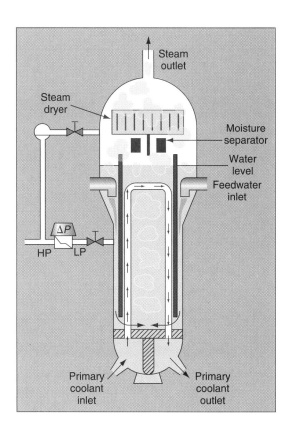

separate pressure detector measures the pressure of the saturated steam. Since saturation pressure is proportional to saturation temperature, a pressure signal can be used to correct the differential pressure for density. An electronic circuit uses the pressure signal to compensate for the difference in density between the reference leg water and the steam generator fluid. As the saturation temperature and pressure increase, the density of the steam generator water will decrease. The ΔP detector should now indicate a higher level, even though the actual ΔP has not changed. The increase in pressure is used to increase the output of the ΔP level detector in proportion to saturation pressure to reflect the change in actual level.

6.22 Environmental Concerns

The density of the fluid whose level is to be measured can have a large effect on level detection instrumentation. It primarily affects level-sensing instruments that utilize a wet reference leg. In these instruments, it is possible for the reference leg temperature to be different from the temperature of the fluid whose level is to be measured. An example of this is the level detection instrumentation for a boiler steam drum (Figure 6–21).

FIGURE 6–21 Drum level differential pressure (notice steam pot needed to read level).

The water in the reference leg is at a lower temperature than the water in the steam drum. Therefore, it is more dense, and must be compensated for to ensure the indicated steam drum level is indicated accurately. Ambient temperature variations will affect the accuracy and reliability of level detection instrumentation. Variations in ambient temperature can directly affect the resistance of components in the instrumentation circuitry, and, therefore, affect the calibration of electric/electronic equipment. The effects of temperature variations are reduced by the design of the circuitry and by maintaining the level detection instrumentation in the proper environment. The presence of humidity will also affect most electrical equipment, especially electronic equipment. High humidity causes moisture to collect on the equipment, causing short circuits, grounds, and corrosion, which, in turn, may damage components. The effects due to humidity are controlled by maintaining the equipment in the proper environment.

FIGURE 6–22 Magnetic float level in foreground; condensate receiver tank level.

■ SUMMARY

Level measurement uses physical properties involving pressure, Archimedes' law, conductivity, resistance, and many other physical relationships to determine accurate levels. These devices require that the installer adequately understand the principles involved and be able to determine if the installation and operation of the device are affected. This chapter provided the physical concerns, the installation requirements, and the proper physical parameters that confront each measurement application.

■ REVIEW QUESTIONS

1. Explain the differences between inferred measurement and direct measurement applications.

2. What equalization force is used by the gauge glass to provide a visual level that is readily observed?

3. Gauge glass applications may be used up to what operating pressure and temperature?

4. Is the ball float a direct or indirect method of measuring level?

5. Determine the calibration span for a differential pressure transmitter that is to measure 10 feet of liquid level with a specific gravity of 1.0.

6. Determine the calibration span for a differential pressure transmitter that is to measure 12 feet of liquid level with a specific gravity of 0.95.

7. Why would a differential pressure sensor measure negative values for an increasing level?

8. One PSI is equal to how many inches of "head"?

9. How much does 1 cubic foot of water weigh?

10. What is the type of error if a transmitter has an excess, or unwanted, pressure on the "HI" side of a differential pressure transmitter? How is it corrected?

11. What concern(s) provides the impetus for density compensation when measuring level?

chapter 7

Fundamentals of Temperature

■ OUTLINE

◼ OVERVIEW

Temperature measurement is unique in that all temperature measurements must be derived from an indirect form. That is, by measuring temperature's effects on some material we can determine the actual temperature that is dependent upon the temperature standard employed. As you would anticipate, there are several ways to introduce errors into a temperature measurement loop. This chapter presents the most commonly used devices in industry today and explains their function, operation, and potential errors. Once the physical parameters of temperature measurement are understood, the art of calibration can begin.

◼ OBJECTIVES

After completing this chapter, you should be able to:

- Describe the purpose of temperature measurement.
- Explain the operation of temperature measurement devices.
- List the four temperature scales used in temperature measurement.
- Explain the reason for a temperature measurement standard.
- Describe the characteristics and operation of a thermocouple.
- Describe the circuit requirements and compensations needed for a thermocouple circuit.
- Explain the operation of an RTD measurement circuit.
- Describe the circuit requirements and operation of an RTD measurement circuit.
- List possible causes of RTD errors.
- Describe how a thermistor measures temperature indirectly.
- Explain how an integrated circuit sensor measures temperature indirectly.

◼ INTRODUCTION

7.1 Temperature Fundamentals

Industries require their energy management systems to perform as the most efficient systems available. When these industries control their temperature processes efficiently, they produce their finished products at a reduced expense. With the demand for natural resources becoming greater, industries today are placing a greater emphasis on energy management systems, resulting in an increased demand for control systems technicians. One of the most important methods to save energy is to monitor and control all heat-related processes. Since heat is energy, the quantity of energy saved has a direct result on profits. As industries place a greater emphasis on the importance of temperature measurement, it follows that the installation and understanding of the fundamentals of temperature measurement will grow in importance for those people who work directly with temperature sensors.

7.2 Common Temperature Measurement Applications

The four most popular types of temperature sensors we will study are: thermocouples, RTDs, integrated circuit sensors, and thermistors. It is important to understand the basics of temperature measurement because, regardless of the device used, the response of temperature elements will remain the same. The only difference will be the temperature scales that record and measure the data.

7.3 Inferred Measurement

The first and most important fact to remember is that you cannot measure temperature directly. The implication of this statement has greater significance than, perhaps, many people realize. We can only measure temperature by observing changes in other objects or materials. The typical thermometer is calibrated in degrees Fahrenheit and we observe temperature change by observing the effect temperature has on mercury. As the temperature increases or decreases, the mercury changes with respect to the temperature change. We observe the change in mercury, not the change of temperature. Temperature will cause all solids, liquids, and gases to expand or contract with temperature change. What is helpful about this principle is that all expand at a uniform rate that can be measured and converted into a temperature scale.

7.4 History of Temperature Measurement

As we have studied in pressure measurement, there are different scales that measure temperature. Fahrenheit is probably the most common in the United States, and Celsius is probably the next most important temperature scale. Other temperature scales are, perhaps, more useful because they record temperature in absolute scales and it is sometimes helpful to think of the absolute temperature scales beginning at a zero temperature the same way a zero absolute pressure reading is obtained from zero pressure. Regardless of the technique or scale used, the important thing to remember is that temperature is the measurement of the presence of heat or energy. There can be as many as four scales used that we need to become familiar with: Fahrenheit, Celsius, Kelvin, and Rankine.

Galileo is credited with inventing the thermometer around the year 1592. He observed a container filled with alcohol and in which a long, narrow glass tube with a reservoir at the upper end was placed. As temperature was increased, the air trapped in the reservoir was also heated and the air was forced out of the tube in which it was trapped. As the temperature was decreased, the remaining trapped air contracted and allowed some of the alcohol to begin advancing up the hollow glass tube. This "upside-down" form of a thermometer was the first thermometer that could be produced in quantity, and the first that recorded the same results when subjected to the same temperature change. Over the years, many different forms of temperature measurement scales were used, but it was not until Daniel Fahrenheit established a scale with zero degrees being the lowest temperature he could record with his mixture of ice water and salt (ammonium chloride). He chose the temperature point of the human body for the upper point. Why 96° and not 100°? It is theorized that Fahrenheit chose the multiple of twelve that earlier measurement scales had been using. Fahrenheit's temperature scale grew in popularity due to its ease of production. Around Farenheit's time, the practice of science increased dramatically, resulting in an increase of scientific data recordings. Scientists would come to recognize the need for temperature scales with alternate units of measurement.

7.5 Temperature Measurement Units

Approximately half a century later, Anders Celsius proposed a temperature scale that contained 100° between water's freezing and boiling points. This was the beginning of the Celsius scale.

Around 1800, Lord Kelvin proposed a universal thermodynamic scale based on the coefficient expansion of an ideal gas. This scientific jargon, put simply, means that Kelvin had created a model that could theoretically establish a point of **absolute zero,** or the absence of heat. He used the Celsius temperature scale to record the changes that his working model could produce. The resulting **Kelvin** temperature scale uses the concept of 100° between water's freezing and boiling points.

The **Rankine** scale is merely another form of measuring absolute temperature using the Fahrenheit scale.

Figure 7–1 shows the relationship among the four temperature scales. Following are some conversion equations for the four temperature measurement units shown in Figure 7–1:

$$C = 5/9 \ (F - 32)$$
$$F = 9/5 \ C + 32$$
$$K = C + 273.15$$
$$R = F + 459.67$$

FIGURE 7–1 Four temperature measurement units.

Fahrenheit	Rankine	Celsius	Kelvin	
212	672	100	373	Boiling point of water
32	492	0	273	Freezing point of water
0	460			
−460	0	−273	0	Absolute zero

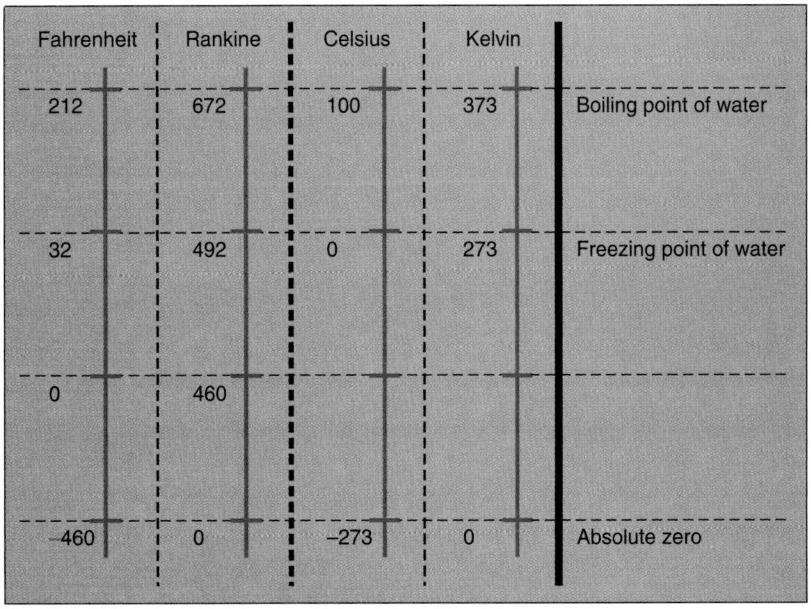

7.6 Temperature Measurement Standard

As stated before, the four most common types of temperature trans-ducers are thermocouples, RTDs, integrated circuit sensors, and ther-mistors. We cannot build a temperature divider or add temperature as we can with resistive elements so we must use temperature responses that have been established by physical properties. The International Practical Temperature Scale (IPTS), a reference scale based on these properties, establishes a reference point. Instruments we use today use these points as a reference to extract temperature readings between these points. Some of these instruments can be fairly exotic, but we will concentrate on the four basic types.

7.7 Purpose of Temperature Measurement

Although the temperatures that are monitored vary slightly depending on the details of facility design, temperature detectors are used to pro-vide three basic functions: indication, alarm, and control. The temper-atures monitored may normally be displayed in a central location, such as a control room, and may have audible and visual alarms asso-ciated with them when specified, preset limits are exceeded. These temperatures may have control functions associated with them so that equipment is started or stopped to support a given temperature condi-tion or so that a protective action occurs.

In the event that key temperature-sensing instruments become in-operative, several alternate methods can be used. Some applications utilize installed spare temperature detectors or dual-element **resist-ance temperature detectors (RTDs)**. The dual-element RTD has two sensing elements; only one is normally connected. If the operating el-ement becomes faulty, the second element can be used to provide tem-perature indication. If an installed spare is not utilized, a contact pyrometer (portable thermocouple) can be used to obtain temperature readings on those pieces of equipment or systems that are accessible.

◾ TEMPERATURE SENSORS

7.8 The Thermocouple

When two dissimilar metals are joined at both ends and one end is heated, a continuous current flows in the circuit. If this circuit is bro-ken at one end, the resulting open circuit voltage is a predictable read-ing that will depend upon the temperature and the metal types used. All dissimilar metals present this effect. By knowing the combination of metals used, we can measure the resulting temperature from the voltage reading (Figure 7–2). The performance of a **thermocouple** ma-terial is generally determined by using that material with platinum. The most important factor to be considered when selecting a pair of materials is the *thermoelectric difference* between the two materials. A significant difference between the two materials will result in better thermocouple performance.

Figure 7–2 illustrates the characteristics of the more common mate-rials used with platinum. Other materials can be used in addition to

FIGURE 7–2 Thermocouple characteristics.

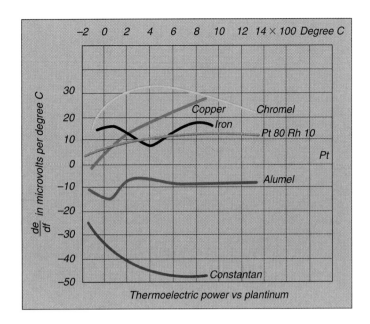

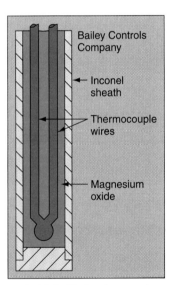

FIGURE 7–3 Internal construction of a typical thermocouple.

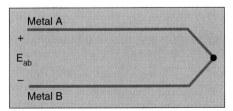

FIGURE 7–4 Example of thermocouple metals and polarity.

those shown in Figure 7–2. For example: Chromel-constantan is excellent for temperatures up to 2,000°F; nickel/nickel-molybdenum sometimes replaces chromel-alumel; and tungsten-rhenium is used for temperatures up to 5,000°F. Some combinations used for specialized applications are chromel-white gold, molybdenum-tungsten, tungsten-iridium, and iridium/iridium-rhodium.

Figure 7–3 shows the internal construction of a typical thermocouple. The leads of the thermocouple are encased in a rigid metal sheath. The measuring junction is normally formed at the bottom of the thermocouple housing. Magnesium oxide surrounds the thermocouple wires to prevent vibration that could damage the fine wires and enhances heat transfer between the measuring junction and the medium surrounding the thermocouple.

In a physical relationship, for any small change in temperature the voltage is linearly proportional. This simply means that as the temperature rises, the voltage will rise at the same rate. We cannot measure the voltage at the open end of the circuit (Figure 7–4) because our voltmeter leads are also made of metal and introduce another form of current.

One way to measure the voltage at junction E_{ab} is to place the thermocouple junction into an ice bath and force the output to be 0°C or 32°F. It is important to understand that the voltmeter reading at 0°C will not be zero because the voltage is dependent on the absolute temperature. For our purposes here, absolute temperature is the temperature scale in which any reading above zero shows the presence of some heat and, conversely, at zero no heat is present and energy (our measuring voltage) does not exist. In other words, at absolute zero there is no current flowing in the junction metals, but any temperature above zero has some current flowing. By forcing the temperature reading of the ice bath to read "zero," we cancel out the voltage that is present below zero.

Thermocouples will cause an electric current to flow in the attached circuit when subjected to changes in temperature. The amount

FIGURE 7–5 Thermocouple circuit.

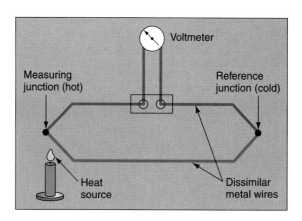

of current that will be produced is dependent on the temperature difference between the measurement and reference junction, the characteristics of the two metals used, and the characteristics of the attached circuit. Figure 7–5 illustrates a simple thermocouple circuit.

Heating the measuring junction of the thermocouple produces a voltage that is greater than the voltage across the reference junction. The difference between the two voltages is proportional to the difference in temperature and can be measured on the voltmeter (in millivolts). For ease of operator use, some voltmeters are set up to read out directly in temperature through use of electronic circuity.

Other applications provide only the millivolt readout. To convert the millivolt reading to its corresponding temperature, you must refer to tables like the one shown in Figure 7–6. These tables can be obtained from the thermocouple manufacturer, and they list the specific temperature corresponding to a series of millivolt readings.

In industry, the thermocouple is used to accurately measure temperatures for indication, alarm, and control. A thermocouple can be subjected to a temperature and its output can be checked to reference sheets to establish a temperature, or the ice bath method can be used. Since the induced emf (output voltage) is often in the range of millivolts, sensitive equipment is used to detect this voltage and convert to temperature.

One of the useful properties that is used effectively when employing thermocouples is an isothermal block. This allows the use of dissimilar metals that will not cause an induced voltage because the temperature will remain the same across both junctions (Figure 7–7). This allows the use of less-expensive lead wires to the recording device since thermocouple wires are much more expensive. Copper wire is often used for thermocouple extension wiring, but it is still important to realize that a temperature difference from the isothermal block will cause an induced emf. For example, the temperature in a control room is hardly ever the same as that in a field junction box. These differences in potential voltage can be corrected as long as the reference temperature of the thermocouple is known. Two of the most widely used thermocouples, iron-constantan and copper-constantan, often use extension leads of the same material. Extension wiring is color-coded to minimize mistakes when connecting wires. Most often, the lead wiring of thermocouples is comparatively small (#14 through #20)

Temperature (°C)(IPTS 1968).											Reference Junction 0°C.	
°C	0	10	20	30	40	50	60	70	80	90	100	°C
				Thermoelectric Voltage in Absolute Millivolts								
-0	0.000	-0.053	-0.103	-0.150	-0.194	-0.236						-0
+0	0.000	0.055	0.113	0.173	0.235	0.299	0.365	0.432	0.502	0.573	0.645	+0
100	0.645	0.719	0.795	0.872	0.950	1.029	1.109	1.190	1.273	1.356	1.440	100
200	1.440	1.525	1.611	1.698	1.785	1.873	1.962	2.051	2.141	2.232	2.323	200
300	2.323	2.414	2.506	2.599	2.692	2.786	2.880	2.974	3.069	3.164	3.260	300
400	3.260	3.356	3.452	3.549	3.645	3.743	3.840	3.938	4.036	4.135	4.234	400
500	4.234	4.333	4.432	4.532	4.632	4.732	4.332	4.933	5.034	5.136	5.237	500
600	5.237	5.339	5.442	5.544	5.648	5.751	5.855	5.960	6.064	6.169	6.274	600
700	6.274	6.380	6.486	6.592	6.699	6.805	6.913	7.020	7.128	7.236	7.345	700
800	7.345	7.454	7.563	7.672	7.782	7.892	8.003	8.114	8.225	8.336	8.448	800
900	8.448	8.560	8.673	8.786	8.899	9.012	9.126	9.240	9.355	9.470	9.585	900
1,000	9.585	9.700	9.816	9.932	10.048	10.165	10.282	10.400	10.517	10.635	10.754	1,000
1,100	10.754	10.872	10.991	11.110	11.229	11.348	11.467	11.587	11.707	11.827	11.947	1,100
1,200	11.947	12.067	12.188	12.308	12.429	12.550	12.671	12.792	12.913	13.034	13.155	1,200
1,300	13.155	13.276	13.397	13.519	13.640	13.761	13.883	14.004	14.125	14.247	14.368	1,300
1,400	14.368	14.489	14.610	14.731	14.852	14.973	15.094	15.215	15.336	15.456	15.576	1,400
1,500	15.576	15.697	15.817	15.937	16.057	16.176	16.296	16.415	16.534	16.653	16.771	1,500
1,600	16.771	16.890	17.008	17.125	17.243	17.360	17.477	17.594	17.711	17.826	17.942	1,600
1,700	17.942	18.058	18.170	18.282	18.394	18.504	18.612					1,700
°C	0	10	20	30	40	50	60	70	80	90	100	°C

FIGURE 7–6 Voltage vs. temperature reference table.

FIGURE 7–7 Calibrating a thermocouple using an ice bath.

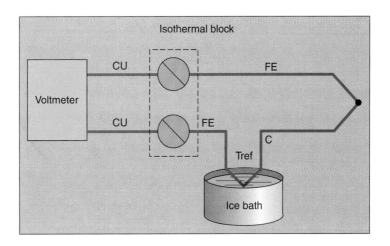

but mostly the size is dependent on the strength required to pull the conductors in a conduit. The resistance of lead wires can also play a part in the calibration of a thermocouple since the resistance will change as lead wiring is added.

A thermocouple that has a known reading for a 0°C temperature can be forced at the control room to read 0°C also. One last factor that is very important to the installation of thermocouples and thermocouple wiring is the location where the thermocouple is mounted and where any resultant wiring is run. Because of the low levels of thermocouple wiring voltages, stray currents or voltages (**noise**) can greatly affect the final reading. Thermocouple lead wiring should always be placed away from high-voltage or high-current conductors. Separate conduit or raceways are always recommended. Shielded twisted-pair wiring is another useful precaution since the shield protects from noise and the twisting of the pair negates additional induced emfs.

Thermocouples are rarely installed directly into the process—they are placed into a protective covering called a **thermowell,** which is placed into the process. This is simpler and protects the thermocouple, which can be removed without stopping the process. Thermowells are available in sizes from ½ inch to 2 inches and are screwed into the vessel or line. Thermowells are often made of 316 or 304 stainless steel, but other alloys are available. Thermowells can slow down the response time of the thermocouple since the thermowell acts as an insulator (although poor), which can isolate the thermocouple from process. Bare, thin strips of like thermocouple metals can be inserted into the thermocouple to reduce the air gap, isolating the thermocouple, and can speed the process with the transfer of heat through the metal parts. Bare or thin-sheathed couples that greatly reduce response times are available, but the prices of these are comparatively greater.

Thermocouple and thermocouple wiring failure, or poor responses, can generally be traced to one of seven possibilities:

1. Poor junction connection
2. Decalibration of thermocouple wire
3. Shunt impedance and galvanic action
4. Thermal shunting
5. Noise and leakage currents
6. Thermocouple specifications
7. Documentation

There are a number of ways to connect thermocouple wiring: soldering, silver-soldering, welding, etc. The important point is that an adequate junction must be made such that any current flowing in the conductor will see a minimal increase in resistance. Soldering a joint limits the temperature at which the wiring can be subjected (solder melts at a relatively low temperature), so welding the joints can allow for higher temperature measurements. With soldering and, especially, welding, care must be taken so that the wiring itself is not degraded from the excessive temperature that is used when making the connection. With any connection point, a degradation of the joint will introduce some added resistance, reducing the output voltage of the thermocouple.

Decalibration can have a more serious effect than an open or shorted circuit because the thermocouple appears to be working normally. Decalibration is the result of altering the thermocouple wiring so that it no longer performs at its rated specifications. Decalibration can be the result of atmospheric diffusion caused by high, extreme temperatures. The most common decalibration occurs when the conductors are "cold-worked" or stressed by drawing the conductors through a conduit. This cold-working stresses and strains the conductor, almost as if it is being stretched; the resulting increase in resistance is introduced into the circuit. Vibration can also play a part in decalibration by degrading the conductor. Annealing can occur in a wire that undergoes constant, large temperature changes.

High temperature can degrade thermocouple insulators. An insulator that is subjected to high temperatures loses its insulating properties. An insulator can be degraded to the point where a "virtual junction" can be created, which is, in fact, a "short" in the circuit. With an insulator that has been broken down, the lack of resistance can create another path for junction current that will result in false readings of the thermocouple; this effect is known as *shunt impedance.*

Galvanic action can change the output of a thermocouple reading. Galvanic action is the chemical breakdown of the metals used in the thermocouple. Sometimes dyes are used in the manufacture of thermocouple insulators. In the presence of water, these dyes will begin a galvanic process that can result in increasing the voltage measurement to many times higher than the original thermocouple output.

Thermal shunting is the process of reducing the effects of temperature on the thermocouple and its associated wiring. As with other devices, thermocouples absorb some of the heat they were intended to measure. If the process to be measured by a thermocouple is relatively small, the resultant thermocouple must also be small. Imagine, for example, a very small volume of a working process. If the thermocouple that will be used to measure the process is large in comparison to the process, the thermocouple will, in fact, draw heat as the process is being heated and contribute heat to the process as the process is being cooled. The result is a "time lag" that will not accurately reflect the temperature of the process. A smaller thermocouple (with smaller wiring) is less likely than a larger thermocouple to have the "time lag" effects placed upon it. To minimize these effects, extension wiring can sometimes be used. Extension wiring is intended to cover long distances between the thermocouple and the voltmeter. This type of wiring is generally larger than thermocouple wiring and is easier to pull into a conduit (over long distances) and is subjected to fewer temperature extremes. A larger extension wire also reduces the effect of resistance on the conductors and is less expensive than thermocouple wiring. Another result of extension wiring is to reduce the amount of noise imposed on the conductor due to the construction of the conductors themselves.

Noise can be imposed on conductors but, with care, the effects can be minimized. As we know, all fluctuating forms of current (ac or fluctuating dc) create a magnetic field that will impose a current on a conductor that is within the field generated. The most important and easiest way to offset this property is to route all field wiring away from possible sources of noise. In some cases, this is not possible, so the contractor can take steps to eliminate the noise by integration, tree switching, and guarding. The controller can cancel the effects of an unwanted reading.

Thermocouple wire is often furnished at a known specification, but the known specification can be checked by calibrating the wire at a known temperature. Thermocouple wire will often resemble thermocouple specifications, but its output voltage may differ slightly and must be "zeroed in." If a wire is calibrated to meet specifications, then all the other possible types of thermocouple error must be considered to avoid decalibration.

With the large number of data points being monitored today, it is possible that a faulty reading on a thermocouple may not be faulty at all. The data points should always be listed on an associated specification sheet and care should always be taken to ensure that these data are recorded accurately.

Supplemental Information

The following information is intended to supplement the thermocouple information discussed earlier. Each of the following types of thermocouples exhibit all of the characteristics discussed earlier, but there are a few distinct differences that can be helpful in eliminating those troublesome loops.

Noble metal thermocouples (types B, R, and S) are all made of platinum or platinum-rhodium metals. High temperatures can cause a form of metallic diffusion that can change the calibration. Therefore, these thermocouples should be used only when installed inside a non-metallic sheath such as high-purity alumina. The only exception is to install these thermocouples inside a platinum sheath, and this option is very expensive. These thermocouples are the most stable of all thermocouples and type S is the most stable of all.

Base metal thermocouples (types E, J, T, K, G, C, D and N) can be used under any chemical composition, unlike the noble metal specifications. Any combination of metals that results in the desired temperature voltage curve can be used. This leads to metal combinations that can be rather exotic (Figure 7–8).

Type E—Ideally suited for low-temperature measurements. Useful for detecting small temperature changes.

Type J—Uses iron for the positive element in a thermocouple. Should never be used above 760°C. Does not follow performance specs especially well due to impurities in the iron. Popular because it is relatively inexpensive.

Type T—Has only one copper lead. Advantage is that the copper material is the same as a digital monitoring device, which makes lead compensation unnecessary.

Type K—To be used at higher temperatures, typically 450°C and above. Alloy content prevents distortions due to oxidation at high temperatures.

Type G,C,D—Normally used in high-temperature areas but not in high-oxidation areas. Made of tungsten-rhenium. They become very brittle above 1,200°C. Rhenium alloys are used in both thermocouple legs to make the wires easier to handle.

Type N—Nickel-based thermocouple system used in very high-temperature measurements. This thermocouple shows great stability at working extremes. Suitable for high-temperature measurement only.

ANSI Code	Alloy Combination		Color Coding		Maximum Temperature Range	EMF (mV) Over Max. Temperature Range	Limits of Error (Whichever is Greater)	
	+ Lead	− Lead	Thermocouple Grade	Extension Grade			Standard	Special
J	Iron Fe (magnetic)	Constantan Copper-nickel Cu-Ni			−32 to 1382°F 0 to 750°C Thermocouple grade 32 to 392°F, 0 to 200°C Extension grade	−8.095 to 69.563	2.2°C or 0.75%	1.1°C or 0.4%
K	Chromega* Nickel-chromium Ni-Cr	Alomega* Nickel-aluminum Ni-Al (magnetic)			−328 to 2282°F −200 to 1250°C Thermocouple grade 32 to 392°F, 0 to 200°C Extension grade	−6.458 to 54.886	2.2°C or 0.75% Above 0°C 2.2°C or 2.0% Below 0°C	1.1°C or 0.4%
V*	Copper Cu	Constantan Copper-nickel Cu-Ni	None established	None established	32 to 176°F 0 to 80°C Extension grade			
T	Copper Cu	Constantan Copper-nickel Cu-Ni			−328 to 662°F −200 to 350°C Thermocouple grade −76 to 212°F, −60 to 100°C Extension grade	−6.258 to 20.872	1.0°C or 0.75% Above 0°C 1.0°C or 1.5% Below 0°C	0.5°C or 0.4%
E	Chromega* Nickel-chromium Ni-cr	Constantan Copper-nickel Cu-Ni			−328 to 1652°F −200 to 900°C Thermocouple grade −32 to 392°F, 0 to 200°C Extension grade	−9.835 to 76.373	1.7°C or 0.5% Above 0°C 1.7°C or 1.0% Below 0°C	1.0°C or 0.4%
N	Omega-P* Microsil Ni-Cr-Si	Omega-n* misil Ni-Si-Mg			−450 to 2372°F −270 to 1300°C Thermocouple grade 32 to 392°F, 0 to 200°C Extension grade	−4.345 to 47.513	2.2°C or 0.75% Above 0°C 2.2°C or 2.0% Below 0°C	1.1°C or 0.4%
R	Platinum-13% rhodium Pt-13% Rh	Platinum Pt	None established		32 to 2642°F 0 to 1450°C Thermocouple grade 32 to 300°F, 0 to 150°C Extension grade	−0.226 to 21.101	1.5°C or 0.25%	0.6°C or 0.1%
S	Platinum-10% rhodium Pt-10% Rh	Platinum Pt	None established		32 to 2642°F 0 to 1450°C Thermocouple grade 32 to 300°F, 0 to 150°C Extension grade	−0.236 to 18.693	1.5°C or 0.25%	0.6°C or 0.1%
U*	Copper Cu	Copper-low nickel Cu-Ni	None established		32 to 122°F 0 to 50°C Extension grade			
B	Platinum-30% rhodium Pt-30% Rh	Platinum-6% rhodium Pt-6% Rh	None established		32 to 3092°F 0 to 1700°C Thermocouple grade 32 to 212°F, 0 to 100°C Extension grade	0 to 13.820	0.5°C over 800°C	Not estalished
G* (W)	Tungsten W	Tungsten-26% rhenium W-26% Re	None established		32 to 4208°F 0 to 2320°C Thermocouple grade 32 to 500°F, 0 to 260°C Extension grade	0 to 38.564	4.5°C to 425°C 1.0% to 2320°C	Not estalished
C* (W5)	Tungsten-5% rhenium W-5% Re	Tungsten-26% rhenium W-26% Re	None established		32 to 4208°F 0 to 2320°C Thermocouple grade 32 to 1600°F, 0 to 870°C Extension grade	0 to 37.066	4.5°C to 425°C 1.0% to 2320°C	Not estalished
D* (W3)	Tungsten-3% rhenium W-3% Re	Tungsten-25% rhenium W-25% Re	None established		32 to 4208°F 0 to 2320°C Thermocouple grade 32 to 500°F, 0 to 260°C Extension grade	0 to 39.506	4.5°C to 425°C 1.0% to 2320°C	Not estalished

FIGURE 7–8   Thermocouple color code and operating ranges.

7.9 The RTD

The resistance of certain metals will change as temperature changes. This characteristic is the basis for the operation of a resistance temperature detector (RTD). The hotness or coldness of a piece of plastic, wood, metal, or other material depends on the molecular activity of the material. Kinetic energy is a measure of the activity of the atoms that make up the molecules of any material. Therefore, temperature is a measure of the kinetic energy of the material in question.

Whether you want to know the temperature of the surrounding air, the water cooling a car's engine, or the components of a nuclear facility, you must have some means to measure the kinetic energy of the material. Most temperature measuring devices use the energy of the material or system they are monitoring to raise (or lower) the kinetic energy of the device. A normal household thermometer is one example. The mercury, or other liquid, in the bulb of the thermometer expands as its kinetic energy is raised. By observing how far the liquid rises in the tube, you can tell the temperature of the measured object. Because temperature is one of the most important parameters of a material, many instruments have been developed to measure it. One type of detector used is the RTD, which is used at many facilities to measure temperatures of the process or materials being monitored.

The RTD incorporates pure metals or certain alloys that increase in resistance as temperature increases and, conversely, those that decrease in resistance as temperature decreases. RTDs act somewhat like an electrical transducer, converting changes in temperature to voltage signals by the measurement of resistance. The metals that are best suited for use as RTD sensors are pure, of uniform quality, are stable within a given range of temperature, and are able to give reproducible resistance-temperature readings. Only a few metals have the properties necessary for use in RTD elements. RTD elements are normally constructed of platinum, copper, or nickel. These metals are best suited for RTD applications because of their linear resistance-temperature characteristics (Figure 7–9), their high coefficient of resistance, and their

FIGURE 7–9 Electrical resistance-temperature curves.

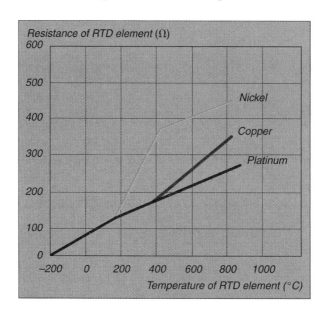

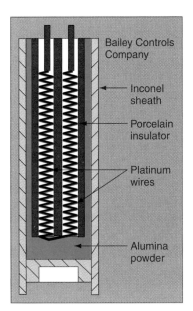

FIGURE 7–10 Internal construction of a typical RTD.

ability to withstand repeated temperature cycles. The coefficient of resistance is the change in resistance per degree change in temperature, usually expressed as a percentage per degree of temperature. The material used must be capable of being drawn into fine wire so that the element can be easily constructed.

RTD elements are usually long, spring-like wires surrounded by an insulator and enclosed in a sheath of metal. Figure 7–10 shows the internal construction of an RTD. This particular design has a platinum element surrounded by a porcelain insulator. The insulator prevents a short circuit between the wire and the metal sheath. Inconel, a nickel-iron-chromium alloy, is normally used in manufacturing the RTD sheath because of its inherent corrosion resistance. When placed in a liquid or gas medium, the inconel sheath quickly reaches the temperature of the medium. The change in temperature causes the platinum wire to heat or cool, resulting in a proportional change in resistance. This change in resistance is then measured by a precision resistance-measuring device that is calibrated to give the proper temperature reading. This device may be a bridge circuit or an electronic transmitter. Both perform the same function (covered in detail later in this text).

Figure 7–11 shows an RTD protective well and terminal head. The well protects the RTD from damage by the gas or liquid being measured. Protecting wells are normally made of stainless steel, carbon steel, inconel, or cast iron, and they are used for temperatures up to 1,100°C.

The common values of resistance for RTDs range from 10 ohms to more than several thousand; the most common is 100 ohms. With the leads connected to an RTD, the total resistance of the circuit may be several ohms more than the RTD. The resistance of leads can present a problem, but the use of a wheatstone bridge can eliminate resistive values that could change the readings of the circuit. The bridge is a practical way to eliminate resistive values since the voltage read is taken directly across the RTD. The bridge will require four connecting wires, three temperature-independent resistors, and a power source. To avoid subjecting the three other resistors to unwanted temperatures, the RTD is connected to the bridge via connecting wires. The bridge operates by placing R_x in the circuit, as shown in Figure 7–12, and then adjusting R_3 so that all current flows through the arms of the bridge circuit.

When this condition exists, there is no current flow through the ammeter, and the bridge is said to be balanced. When the bridge is balanced, the currents through each of the arms are exactly proportional.

FIGURE 7–11 RTD protective well and head.

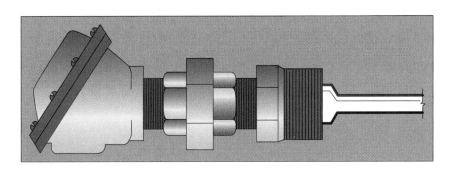

FIGURE 7-12 Unbalanced bridge circuit.

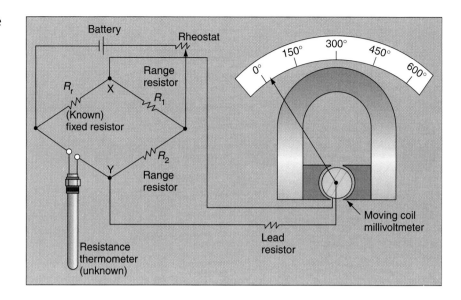

FIGURE 7-12 Unbalanced bridge circuit.

They are equal if $R_3 = R_2$. Most of the time, the bridge is constructed so that $R_1 = R_2$. When this is the case, and the bridge is balanced, then the resistance of R_x is the same as R_3, or $R_x = R_3$. When balance exists, R_3 will be equal to the unknown resistance, even if the voltage source is unstable or is not accurately known. A typical Wheatstone bridge has several dials used to vary the resistance. Once the bridge is balanced, the dials can be read to find the value of R_3. Bridge circuits can be used to measure resistance to tenths or even hundredths of a percent accuracy. When used to measure temperature, some Wheatstone bridges with precision resistors are accurate to about +0.1°F. Two types of bridge circuits (unbalanced and balanced) are utilized in resistance thermometer temperature-detection circuits. The unbalanced bridge circuit (Figure 7–12) uses a millivoltmeter that is calibrated in units of temperature that correspond to the RTD resistance.

The three-wire RTD can alleviate most problems with resistance but, unless you can accurately measure the load resistance or balance the three other bridge resistors, some deviations in accuracy will remain. In some cases, this circuit may be all that is required, but in other cases a more accurate measurement circuit is needed. Consider, then, what is needed to correct our circuit. We need to be able to measure the voltage drop across the RTD without losing any accuracy due to unwanted resistance. If we can measure the voltage without causing a change in the circuit current, we can obtain accurate results. The answer is the four-wire RTD (Figure 7–13).

FIGURE 7-13 Four-wire RTD circuit.

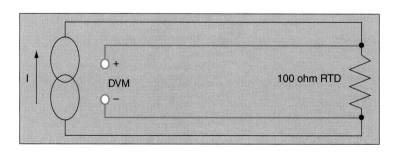

We can control the RTD current by using a current source for the power supply. By adding the two extension wires for the voltage take-off, we can monitor the voltage across the RTD remotely. The addition of the voltmeter does not introduce a significant load change to the circuit because the resistance of the voltmeter is several times higher than the RTD. For all practical purposes, the voltmeter leads can be considered an open circuit with no current flow and, therefore, no voltage drop.

The output voltage is directly proportional to the resistance of the RTD and only one conversion equation is used. The digital voltmeter measures only the voltage across the RTD and is insensitive to the lead length. In fact, the longer the leads are, the better, since this merely will oppose more current flow when the meter is connected. The only disadvantage to using a four-wire RTD is that an extra extension wire needs to be run to the voltmeter.

The same practical precautions that apply to thermocouples must be observed for RTDs. Use shields where possible, use twisted-pair wiring, avoid stressing the conductors, use a larger-size wire for long pulls, avoid high-temperature areas, and maintain adequate documentation.

The RTD is more fragile than a thermocouple and care must be taken when installing it to minimize damage.

The RTD is not self-powered (like the thermocouple). A current must be impressed through the RTD that will, in itself, cause some additional heating. To reduce self-heating effects, use the smallest-size measuring current that will produce the voltage output wanted. Also use the largest-size RTD that will still produce an adequate response time.

The platinum-to-copper connection that is made when the RTD is measured can introduce some additional voltage, although very small. By referencing the resistance readings of the RTD at a known temperature, these effects can be eliminated.

7.10 The Thermistor

The thermistor, like the RTD, is a temperature-sensitive resistor. The thermocouple is the most versatile of temperature sensors and the RTD is the most stable. The thermistor is the most sensitive of all the methods. An additional advantage to the thermistor is that the measurement lead errors that can be introduced are often 500 times less than an equivalent RTD measurement circuit.

The thermistor is made chiefly of semiconductor materials that allow it to be many times more sensitive than thermocouples or RTDs. The disadvantage is that thermistors do not maintain a direct proportional resistance change with temperature (Figure 7–14). The nonlinear curve that determines temperature can "increase" as temperature increases; this is called a positive temperature coefficient thermistor. A negative temperature coefficient implies that as temperature increases, the output of the thermistor decreases.

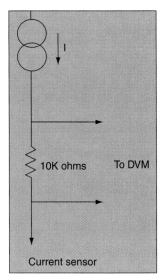

FIGURE 7–14 Current driven thermistor circuit.

The thermistor "works" the same as a RTD but the nonlinearity of the response provides the difference in measuring techniques. A lot of research has focused on the development of a linear thermistor, but today's microprocessor-based controllers provide an easy computing point to calculate temperature/resistance readings. As a rule of thumb, a thermistor is connected to a microprocessor-based controller or voltmeter to establish a baseline of temperature readings for either control or monitoring.

An additional disadvantage is that thermistors are very susceptible to high-temperature damage because they are made of a semiconductor material. They are usually limited to a few hundred degrees Celsius. Continued exposure to high temperature well within the range specified can still lead to **drift.** Thermistors are often very small, which means their response time is fairly fast; this also means that they are susceptible to self-heating errors. Finally, thermistors are even more fragile than thermocouples and RTDs and must be carefully installed to prevent crushing or bond separation (Figure 7–15).

7.11 The Integrated Circuit Sensor

A recent invention in the field of temperature measurement is the use of the integrated circuit transducer. A voltage and current output version can be applied depending on the circuit design and voltage limitations. Both of these models produce an output that is directly proportional to any temperature change. A unique advantage to this circuit is that an analog value for temperature changes can be transmitted to remote locations for monitoring, control, or indication. The same problems will apply to this type of temperature measurement, like all the others, and there are set parameters under which it can be used. Proper documentation is again important to ensure that it is used under the proper specifications.

FIGURE 7–15 Voltage set thermistor.

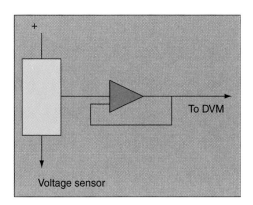

FIGURE 7–16 RTD temperature transmitter.

FIGURE 7–17 Temperature transmitter and well.

FIGURE 7–18 Hot and cold baths for primary elements (sensors).

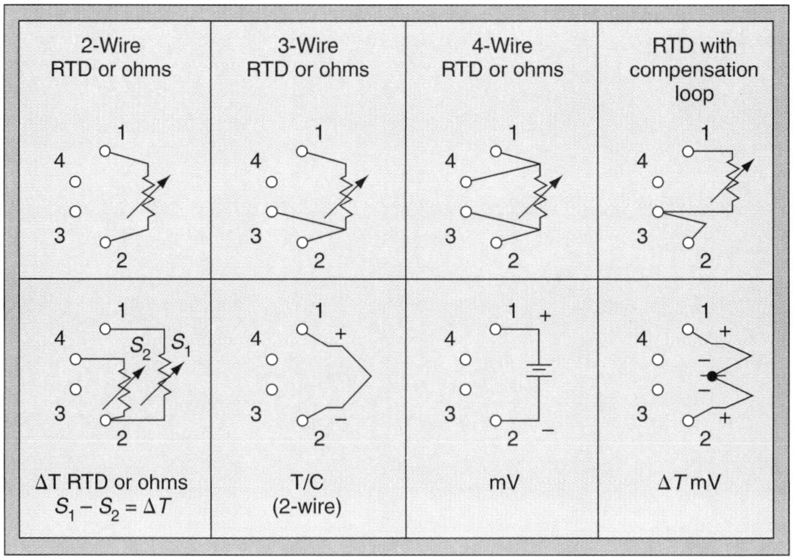

FIGURE 7–19 RTD wiring diagram.

■ SUMMARY

Temperature measurement is a fascinating field wherein numerous conditions and measurement devices can be utilized. Understanding all of the principles of operation, physical properties, circuit layout, and sensor types is a must before accurate calibration of measurement devices can be achieved. This chapter provided you with the information to determine the operation of various devices and their related configuration.

■ REVIEW QUESTIONS

1. List the four common measurement sensors employed in the process industry today.

2. Is temperature a form of direct measurement or indirect measurement? Explain.

3. How many units of measurement are there in the Fahrenheit scale and the Celsius scale from the melting point of ice to the boiling point of water?

4. Determine the equivalent Celsius temperature for 112°F.

5. What is the most important factor to be considered when selecting a pair of materials to perform as a thermocouple?

6. When do thermocouples cause an electric current to flow?

7. Explain the purpose for using standards for calibration of thermocouples.

8. Describe one of the useful properties of the isothermal block used within the thermocouple circuit.

9. What characteristic explains the operation of the RTD?

10. Are RTDs inserted directly into the process measured?

11. Wheatstone bridge-type RTD measurement circuits are known to be how accurate?

12. List some practical precautions given in this chapter that you should observe when you install an RTD measurement circuit.

chapter 8

Fundamentals of Pneumatics and Control Valve Actuators

■ OUTLINE

■ OVERVIEW

Pneumatic principles and control valve actuation provide a variety of working forces used in the controls environment today. Understanding the principles used to provide actuation and indication allows for proper installation and maintenance of control valve actuators, pneumatic supplies, valve position indicators, and general control loop performance. This chapter provides detailed descriptions of the pneumatic devices utilized today and explains their methods of operation.

■ OBJECTIVES

After completing this chapter, you should be able to:

- Describe the purpose and use of pneumatic amplifiers.
- Explain the operation of the nozzle/flapper system.
- Provide a detailed explanation of the operation of an I/P transducer.
- Detail the operational functions of a pneumatic actuator.
- Describe the purpose and operation of positioners.
- Explain the operation of limit, reed, potentiometers, and LVDT valve position indicators.

■ INTRODUCTION

8.1 Fundamentals of Pneumatics and Control Valve Actuators

The term *pneumatic instrumentation* covers a wide range of applications in the instrumentation and controls field where a monitoring or controlling system is needed. Perhaps the most common application is to use the force (pressure) of a gas (air) to move a piston or diaphragm. Pneumatics can also be used as a measuring and signaling medium.

In a pneumatic system, data are carried by the pressure of a gas in a pipe. If a length of pipe is used for a pressure signal, increasing the pressure at one end will cause the pressure to be propagated down the length to the other end, also raising the pressure. This method of conveying information is much slower than the method of electronic control. The change of pressure will travel down the length of the pipe at approximately the speed of sound (1,082 ft/sec). For many process applications, the time delay does not affect the end product.

A pneumatic type of signal was used for many years before electronic devices began to emerge. Pneumatic systems were considered then, and now, to be the safest way to control a process for some applications. Pneumatics are still used today in some applications: as a carryover from the past—it has not been cost effective to remove them, or the threat of sparks from an electrical signal does not warrant their change. For most applications, the gas used is "dry instrument air," or nitrogen for explosive hazardous areas. Signal information usually has been adjusted to a range of 3–15 PSI. There are basically three different forms of pneumatic signal conversions: amplification, nozzle/flapper, and current-to-pressure (I/P).

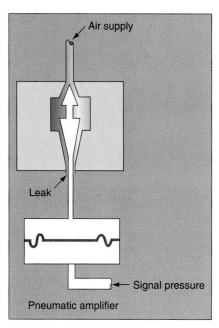

FIGURE 8–1 Two-dimensional view of a pneumatic amplifier.

8.2 Pneumatic Amplifiers

A pneumatic **amplifier** (booster or relay) adjusts the pressure and/or volume by a linear value with respect to the input signal (Figure 8–1). An example of an amplifier would be a "gain" of 10 applied to the standard operating pressure of 3–15 PSI and the result would be a range of 30–150 PSI. **Amplification** is usually accomplished by the use of a regulator. A booster can be either forward- or reverse-acting, depending on the service required. The principles are reversed if an opposite action is wanted. A high signal pressure that will cause the output pressure to decrease is known as a *reverse-acting booster*. Likewise, the lower the input pressure, the higher the output pressure; this is called a forward-acting booster. Although many types of designs are used, the principles remain the same for all. The most important fact is to remember that a pneumatic amplifier converts input signals to a higher pressure, or the same pressure with a greater volume. An amplifier is often used to raise or lower a specific area's controlling pneumatic pressure as needed.

An often-used method to convert mechanical motion to pneumatic pressure and vice versa is to use the nozzle/flapper system. This system usually has a regulated air supply (usually greater than 20 PSI) applied to a singular tube (nozzle) with a restricted area and a "flapper" over the other end of a set tension strength (Figure 8–2). As pressure is increased, the pressure forces the flapper off of its seat on the nozzle. As air pressure is further increased, the nozzle is moved farther away from the nozzle by the force of the applied pressure. When the air pressure is decreased, the flapper returns toward its seat until the force (tension) of the nozzle is greater than the force of the applied pressure. The sensitivity of operation of a nozzle/flapper system is great and it is easy to see that a small adjustment on the flapper can cause a significant change in resulting output pressure. A nozzle/flapper device is often used as a regulating device for a control system that varies its output pressure for a predicted control response.

FIGURE 8–2 Working schematic for a pneumatic controller.

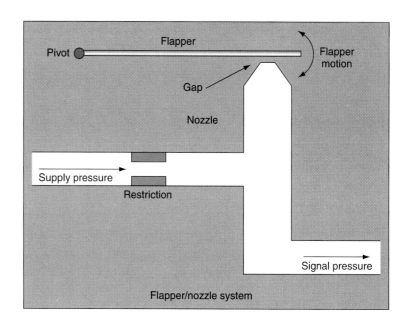

FIGURE 8–3 Schematic of an I/P
transducer.

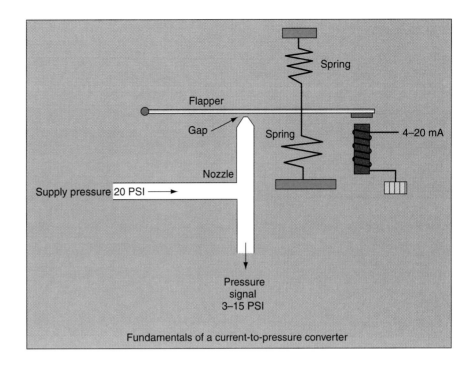

FIGURE 8–3 Schematic of an I/P transducer.

8.3 Current to Pneumatic Transducer (I/P)

Perhaps the most common signal conversion technique used today is the current to pneumatic transducer (**I/P converter**). The result of the I/P converter is the same as that of a nozzle/flapper system—to regulate the output pressure. It is often desirable to regulate the position of a valve by a low current (4–20 mA) over long distances or when a faster response time is needed (Figure 8–3). The I/P converter provides a linear way to position a valve. For a direct-acting transducer, a 4-mA signal would result in a signal pressure that is 0%; a 12-mA signal would translate into a signal pressure of 50%; and a signal of 20 mA would result in a signal pressure of 100%. A reverse-acting transducer contains the same relationship but with the 0% pressure signal established at 20 mA, not 4 mA. With the I/P, the flapper is adjusted by the use of a coil to attract the flapper toward the nozzle, reducing signal pressure. The use of springs and flapper adjustment allow the converter to be adjusted easily. For example, a direct-acting converter would have the adjustment so that 4 mA would equal 3 PSI and 20 mA would correspond to 15 PSI. We should understand by now that the transducer is a method of converting energy of one form to another.

Most controllers today output a signal of the current type (4–20 mA) and most valves today use pneumatic pressure to force valve movement. The transducer converts the current (usually 4–20 mA) to a comparable pneumatic pressure (usually 3–15 PSI). The use of the nozzle flapper arrangement again provides the method of calibration.

Transducers (I/P) require calibration also and the same characteristics of zero and span can be applied. Most transducers allow for field adjustments to be made even though the transducer was probably ordered with the correct working parameters for pneumatics and electric signals calibrated from the factory.

FIGURE 8–4 I/P transducer.

The working components of a device are subject to wear and deformation and therefore will need adjusting. The I/P will require that a specified current can be measured and the output pressure can be regulated by tuning the flapper to adjust the outlet pressure.

The same observations should be made of the transducer as other pneumatic devices (Figure 8–4). The proper venting of bleed pressures, the venting of exhaust pressures, and general working order should be verified. Field technicians who troubleshoot/repair a transducer must have an adequate understanding of how it operates.

A transducer is a very simple form of the nozzle/flapper arrangement. A coil is wound or fixed around a permanent magnet. When a current flows in the coil, the electromatic forces and the permanent magnetic forces are in direct opposition. The opposition of forces causes a flapper movement to occur, forcing the flapper to open or close depending on the construction. A nozzle/flapper can be designed so that it opens or closes when current is applied.

You can perform a simple self-test using working pressures and currents to determine an I/P's design. Connect a current supply to the transducer and run through the range of the input signal (usually 4–20 mA) and record the output pressures of the transducer. You should be able to determine if the transducer is direct-acting, follows 3–15 PSI from 4–20 mA, or reverse-acting, 15–3 PSI from 4–20 mA. That is the crux of the maintenance problems of an I/P. If the construction type of the I/P is known, the I/P can be adjusted to output 3 PSI for 4 milliamps or 15 PSI for 4 milliamps for reverse acting. Reversing of a transducer is accomplished by switching the input leads and recalibrating for the desired range.

Most transducers include a booster (sometimes called a *pneumatic relay*) that allows for the boosting of the pneumatic pressure. These boosters, the same as the relays we have studied before, are capable of amplification. By using boosters, a transducer is available with a number of input and output ranges for various configuration properties such as split ranging.

Split range systems are designed such that a control valve may perform from the 3–9 PSI range while another may operate from the 9–15 PSI range. Split range designs provide a wider range of control that, perhaps, cannot be controlled by a single valve. The calibration of the transducer makes this possible. The versatility of the transducer is evident in its ease of calibration.

A transducer can be calibrated over a wide range, including split ranging, but it is also reversible. Again, working pressures are involved and care must be taken when disassembling transducers and when performing other types of transducer maintenance procedures.

8.4 Pneumatic Actuators

Another pneumatic principle we must become familiar with is the pneumatic actuator. The **actuator** receives a control signal and translates the signal into a force or torque action as it is required to position a final control element. The concept of the actuator is based on the principle of pressure as a force per unit area. A **diaphragm** of a known surface area

FIGURE 8–5 Pneumatic actuator, air-to-close, spring-to-open.

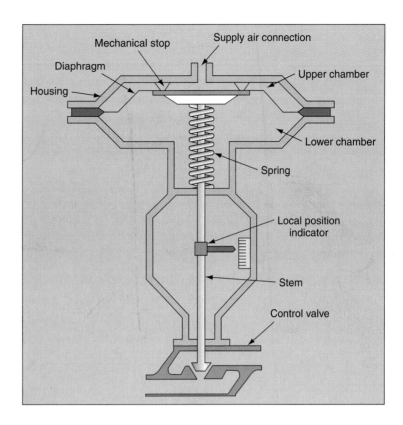

can have a force applied to it by differential pressure. Often, the diaphragm casing is vented on one side, so the differential pressure across a diaphragm is a gauge pressure applied to the inlet pressure port.

The working action of an actuator is a simple conversion of pressure to force (Figure 8–5). As pressure is applied to one side of the diaphragm, the diaphragm is distorted away from the pressure and moves an attached actuator arm that is connected to a valve or other appendage. Springs are used to set a resistance torque and are linearly related to the applied force. In other words, the compression of the spring is set to restrict the amount of diaphragm movement for a specific input pressure. This is why to "bench-set" a valve, the valve body is de-coupled from the valve actuator. Valve actuator stem movement is then set so that a full range of motion is obtained when the actuator is given "zero" and "span" input pressures. For example, the I/P supplying the actuator can be calibrated for a 3–15 PSI span, but the valve can be bench-set to operate on a 5–15 PSI span. The spring has another important function: to return the valve to a set position on a loss of signal pressure. This is known as the *fail-safe* valve position.

An important safety feature is provided by the spring in an actuator. It can be designed to position a control valve in a safe position if a loss of supply air occurs. On a loss of supply air, the actuator shown in Figure 8–5 will fail-open. This type of arrangement is referred to as "air-to-close, spring-to-open" or, simply, "fail-open." Some valves fail in the closed position. This type of actuator is referred to as "air-to-open, spring-to-close," or "fail-closed." This "fail-safe" concept is an important consideration in control safety design.

The signal from a controller is interpreted by the transducer and the corresponding pressure is applied to the diaphragm and the valve stem changes valve position accordingly. Valve-packing friction, process pressure, and hysteresis from actuator response can cause an incorrect positioning of the valve. When errors are introduced in the valve positioning, a positioner can be used to alleviate most of the errors.

8.5 Positioners

In a closed-loop system, the goal is to keep one or more process variables within tolerances specified by the controller. The small tolerances for error sometimes require the use of a positioner to "position" the valve as it is signaled to be placed.

A positioner is a device that regulates the supply air pressure to a pneumatic actuator. It does this by comparing the actuator's demanded position with the control valve's actual position. The demanded position is transmitted by a pneumatic or electrical control signal from a controller to the positioner. The pneumatic actuator in Figure 8–5 is shown in Figure 8–6 with a controller and positioner added. The controller generates an output signal that represents the demanded position. This signal is sent to the positioner. Externally, the positioner consists of an input connection for the control signal, a supply air input connection, a supply air output connection, a supply air vent connection, and a feedback linkage. Internally, it contains an intricate network of electrical transducers, air lines, valves, linkages, and necessary adjustments. Other positioners can also provide controls for lo-

FIGURE 8–6 Pneumatic actuator with controller and pneumatic positioner.

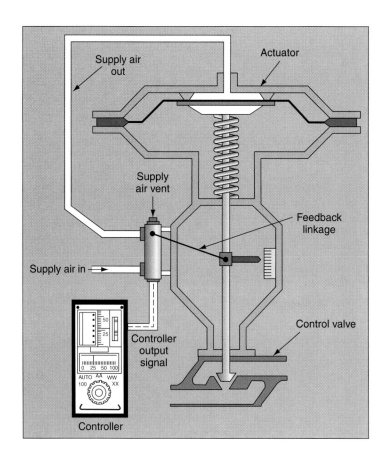

cal valve positioning and gauges to indicate supply air pressure and control air pressure (for pneumatic controllers).

For a direct-acting actuator example, as the control signal increases, a valve inside the positioner admits more supply air to the actuator. As a result, the control valve moves downward. The linkage transmits the valve position information back to the positioner. This forms a small, internal feedback loop for the actuator. When the valve reaches the position that correlates to the control signal, the linkage stops supply air flow to the actuator. This causes the actuator to stop. On the other hand, if the control signal decreases, another valve inside the positioner opens and allows the supply air pressure to decrease by venting the supply air. This causes the valve to move upward and open. When the valve has opened to the proper position, the positioner stops venting air from the actuator and stops movement of the control valve.

A positioner is also used when accurate response is demanded quickly for a process variable. The positioner receives its input from the controller and determines from the controller's signal what the desired valve position should be. A feedback assembly is mounted on the valve stem that provides feedback to the positioner, which is used to compare the actual stem position to the desired position (Figure 8–7).

FIGURE 8–7 Control valve with pneumatic actuator and smart positioner.

If corrections are needed, the positioner calls for an actuator adjustment until the desired and actual positions are the same. The comparison of positions is done mechanically and the output is a pneumatic pressure.

The pressure sources used by a positioner (control) are often separate from the pressure that is used to force diaphragm movement (force). The different pressure sources allow for a positioner that is using the conventional pneumatic range of 3–15 PSI to control an alternate pressure, such as a 6–30 PSI range.

To calibrate a positioner we must recall the concepts of zero and span. A positioner is calibrated by the same method as we have studied earlier, but before calibration starts, all of the relevant information has to be gathered. Signal pressure (example: 3–15 PSI), full or split range (for example 3–9 PSI and/or 9–15 PSI), supply pressure, actuator working pressure range, valve travel, and knowing if the action is forward or reverse acting must be determined before the necessary calibration procedure can begin. The calibration of a positioner has to assume that the actuator is adjusted as it should be within its proper working range (bench set).

The objective with positioner calibration is to establish an accurate starting point referred to as *zero*. The controller is manipulated to send an equivalent zero signal (4 mA for electronic signal, 3 PSI for pneumatic) and the positioner compares the *controller's* signal to the valve position. The positioner is adjusted such that at the zero signal, the positioner is calling for a zero position, full open for a direct-acting, air-to-close valve. As the signal is increased, the positioner should be adjusted (if needed) to call for an increase in the actuator working pressure. Therefore, the two variables that need to be monitored are the controller's signal and the valve stem position.

Adjust the zero setting to allow the stem position to remain at zero and, accordingly, throughout the valve stem range. If the valve has a linear response, the valve stem position should be at 50% for a 50% controller signal. The span adjustment is adjusted until the span of the valve corresponds with the controller signal. The valve stem should indicate full travel when 100% of controller signal is received.

When setting the span for valve positions, it is important that the span adjust is not made to where it takes an exaggerated signal to move the valve from its furthermost point of travel. Some manufacturers suggest that the valve travel for 100% should be set at a value slightly less than full range. Likewise, the valve should be adjusted fully closed when a signal slightly greater than zero position is received. This is to ensure that valve full range of travel is set given full signal ranges.

There are reverse-acting positioners and actuators; therefore, it is necessary to obtain the proper "specs" for each assembly before proceeding. It is also essential that, when adjusting for span in the positioner adjustment, zero is referenced again since the two do interact.

A positioner may also be of the "smart" type, which requires the use of a communicator to calibrate. A smart positioner will have to be set at zero and span, but will also include a setting for midpoint (50%). An adjustable feedback linkage will have to be checked or set to provide accurate indications of valve stem travel.

Pneumatic instruments, devices, and transducers all contain mechanical linkages and appendages. With mechanical parts, ambient conditions will play a greater role in determining if a device is functioning correctly or will function correctly when installed. Ambient compensation may be needed to adjust a transducer that is calibrated in a controlled environment but is installed in a working environment. The dead band of pneumatic instruments will probably be more noticeable. Hysteresis will be a consideration when calibrating an instrument. The mounting locations of pneumatic devices will have to be considered also. A pneumatic device needs to be placed away from areas with large temperature swings, if possible.

A hazardous location for a pneumatic instrument may not be as great a concern as the location of an electronic instrument, but a "dirty" environment will play a larger role in creating control errors, whether it is a calibration problem or fouling of a moving linkage. When calibrating a pneumatic instrument, care must be taken to calibrate the device in a simulated environment as close as possible to the environment of the location where it will be mounted. The mechanical parts in the device are subject to gravitational forces and a device that is installed differently than calibrated will have a tendency to drift toward the high- or low-range limits. Pneumatic instrumentation is a proven monitoring and controlling means that has been in use much longer than electronics. The same zero shift and nonlinearity errors will still be present and are calibrated and adjusted by the same methods as those for an electronic instrument (Figure 8–8). The fundamentals of pneumatics must

FIGURE 8–8 Kinetrol control valve, internal components; note zero and span adjustments.

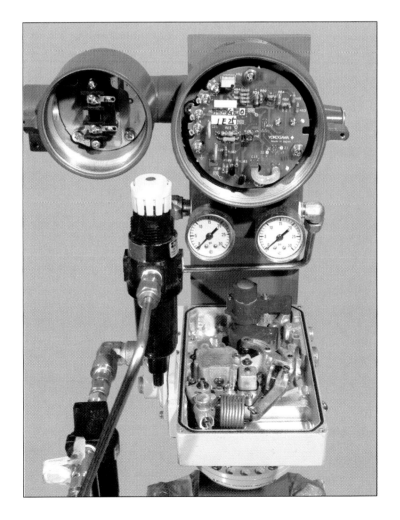

FIGURE 8–9 Limit switches.

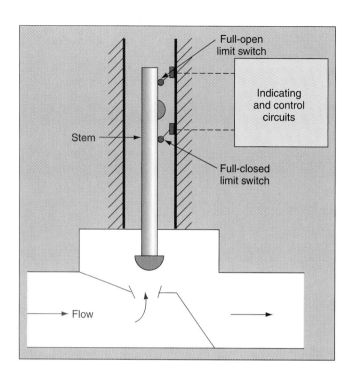

be understood to calibrate final control elements. Air-actuated control valves are and will probably continue to be the only elements with a "force" strong enough to position.

8.6 Limit Switches

A limit switch is a mechanical device that can be used to determine the physical position of equipment. For example, an extension on a valve shaft mechanically trips a limit switch as it moves from open to shut or shut to open. The limit switch gives ON/OFF output that corresponds to valve position. Normally, limit switches are used to provide full-open or full-shut indications, as illustrated in Figure 8–9.

Many limit switches are the push-button variety. When the valve extension comes in contact with the limit switch, the switch depresses to complete, or turn on, the electrical circuit. As the valve extension moves away from the limit switches, spring pressure opens the switch, turning off the circuit.

Limit switch failures are normally mechanical in nature. If the proper indication or control function is not achieved, the limit switch is probably faulty. In this case, local position indication should be used to verify equipment position.

8.7 Reed Switches

Reed switches, illustrated in Figure 8–10, are more reliable than limit switches due to their simplified construction. The switches are constructed of flexible ferrous strips (reeds) and are placed near the intended travel of the valve stem or control rod extension.

FIGURE 8–10 Reed switches.

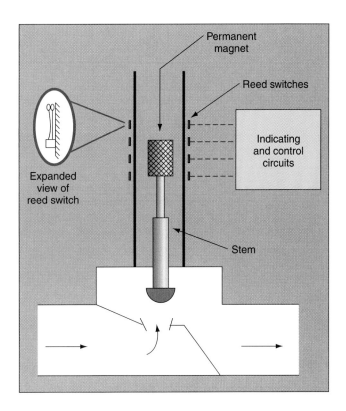

When using reed switches, the extension used is a permanent magnet. As the magnet approaches the reed switch, the switch shuts. When the magnet moves away, the reed switch opens. This ON/OFF indicator is similar to mechanical limit switches. By using a large number of magnetic reed switches, incremental position can be measured. This technique is sometimes used in monitoring a control valve's stem position.

Failures are normally limited to a reed switch that is stuck open or closed. If a reed switch is stuck shut, the open (closed) indication will be continuously illuminated. If a reed switch is stuck open, the position indication for that switch remains extinguished regardless of valve position.

8.8 Potentiometer Position Indicators

Potentiometer valve position indicators (Figure 8–11) provide an accurate indication of position throughout the travel of a valve or control rod. The extension is physically attached to a variable resistor. As the extension moves up or down, the resistance of the attached circuit changes, changing the amount of current flow in the circuit. The amount of current is proportional to the valve position.

Potentiometer valve position indicator failures are normally electrical in nature. An electrical short or open will cause the indication to fail at one extreme or the other. If an increase or decrease in the potentiometer resistance occurs, erratic indicated valve position occurs.

FIGURE 8–11 Potentiometer valve position indication.

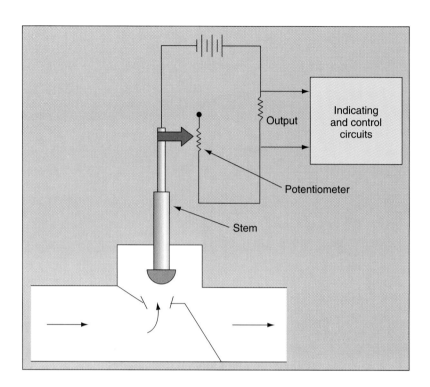

8.9 Linear Variable Differential Transformer (LVDT)

A linear variable differential transformer (LVDT) is a device that provides accurate position indication throughout the range of valve or control rod travel. Unlike the potentiometer position indicator, no physical connection to the extension is required. The extension valve shaft, or control rod, is made of a metal suitable for acting as the movable core of a transformer. Moving the extension between the primary and secondary windings of a transformer causes the inductance between the two windings to vary, thereby varying the output voltage proportional to the position of the valve or control rod extension.

Figure 8–12 illustrates a valve whose position is indicated by an LVDT. If the open and shut position is all that is desired, two small, secondary coils can be utilized at each end of the extension's travel. LVDTs are extremely reliable. As a rule, failures are limited to rare electrical faults that cause erratic or erroneous indications. An open primary winding will cause the indication to fail to some predetermined value equal to zero differential voltage. This normally corresponds to midstroke of the valve. A failure of either secondary winding will cause the output to indicate either full open or full closed.

FIGURE 8-12 Linear variable differential transformer (LVDT).

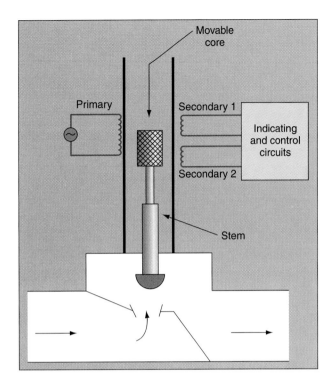

FIGURE 8-13 Actuator symbols.

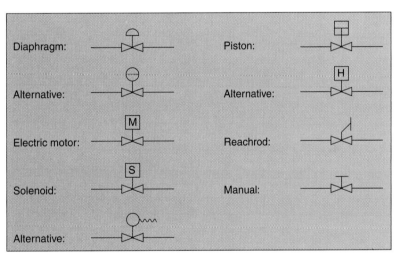

SUMMARY

Pneumatic actuation of control valves is a simple process. The determining factor for correct operation depends on the proper installation and calibration of the devices that provide actuation, feedback, position, and indication of position. This chapter has provided the schematics, wiring, and configuration parameters to allow proper installation of these devices. This chapter also provided the proper operation of the final control element assembly to allow for proper interpretation of control loop performance.

■ REVIEW QUESTIONS

1. Explain the operation of an I/P.

2. Determine the output signal of an I/P with an input range of 4–20 mA and an output range of 3–15 PSI, if an input of 7 milliamps was received by the I/P.

3. List the operating ranges for a direct-acting actuator with a current signal of 4–20 mA and a pneumatic pressure range of 3–15 PSI.

4. To calibrate an I/P to be reverse acting, what physical change must be applied?

5. Explain the designed operating function of split range control and give the operating ranges of the transducers required to control split range applications.

6. An air-operated actuator that is set up to be air-to-close can be referred to by its fail position. Give the fail position for the actuator.

7. Is bench-set performed with the valve body coupled to the actuator or de-coupled?

8. Limit switches provide what information in respect to control valve performance?

chapter 9

Fundamentals of Controllers

■ OUTLINE

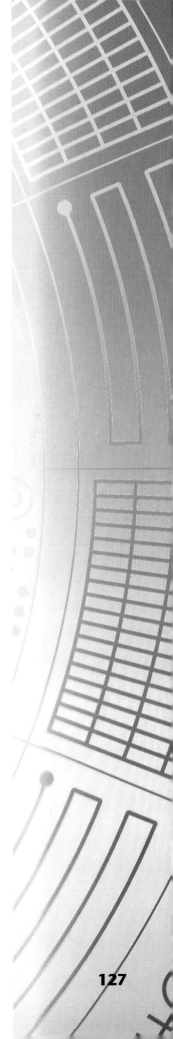

■ OVERVIEW

Automatic controllers are used in a variety of applications. Batch control, process control, and emergency shutdown systems are examples of automatic control systems. This chapter describes the purpose of automatic control, I/O signal types, the components of a microprocessor, and the general functionality of automatic controllers. This chapter does not cover in detail all of the control logic systems, I/O assignments, etc., but rather tries to provide information so that someone who installs and calibrates field I/O does not view the controller as a "black box-type" system.

■ OBJECTIVES

After completing this chapter, you should be able to:

- Compare the concepts of automatic control and manual control.
- Describe the field I/O used by a controller.
- List the signal types received by a controller.
- List the signal types used by a controller to control a process.
- Describe a microprocessor's components and give their respective functions.
- Explain the importance of field I/O performing data gathering and process control.

■ INTRODUCTION

9.1 Fundamentals of Controllers

Instrumentation provides the various measurements and indications used to operate a facility. In some cases, operators record these indications for use in day-to-day operation of the facility. The information recorded helps the operator evaluate the current condition of the system and take actions if the conditions are not as expected.

Requiring the operator to take all of the required corrective actions is impractical, or sometimes impossible, especially if a large number of indications must be monitored. For this reason, most systems are controlled automatically once they are operating under normal conditions. Automatic controls greatly reduce the burden on operators and make their job manageable. Process variables that require control in a system include, but are not limited to, flow, level, temperature, and pressure. Some systems do not require all of their process variables to be controlled—think of a central heating system.

9.2 Automatic Control Concept

A basic heating system operates on temperature and disregards the other atmospheric parameters of the house. The thermostat monitors the temperature of the house. When the temperature drops to the value selected by the occupants of the house, the system activates to raise the temperature of the house. When the temperature reaches the desired value, the system turns off. Automatic control systems neither replace nor relieve the operator of the responsibility for maintaining the facility. The operation of the control systems is checked periodically to verify proper operation. If a control system fails, the operator must be able to take over and control the process manually. In most cases, understanding how the control system works aids the operator in determining if the system is operating properly and in choosing which actions are required to maintain the system in a safe condition.

In the process environment of yesterday, field operators used a slide rule to help them perform calculations to determine the process mixtures, valve alignments, etc. The slide rule could not calculate any problem without interface or direction from the operator; it had to be directed to provide solutions. The process environment of today still calls for operators to predict what will happen to their process given any circumstance, but the control of the process is often dictated by a microprocessor. Still, a computer cannot perform functions that it has not been programmed to initiate. A computer cannot be a standalone controller—it must prompt for and receive guidance from an operator or receive all its inputs from the field instruments. Field technicians who adequately understand their job must be familiar with the operation of a microprocessor-based controller and know how their instruments interact with it.

9.3 Microprocessor-Based Control

Computers are designed so that they will perform only one task at a time even though that task is often accomplished in a few microseconds. The "brain" of a computer does not function as a brain at all—it can only follow a detailed set of instructions. The instructions are part of the "software," or the program that will, for most instances, determine an output given any set of inputs that the program can use. It is easy to see that the program for one process may not be of any use except for the application for which it was written. The program for a blending station will not perform at all when "downloaded" to the microprocessor for a boiler. With software, the application is determined by the process, but there are some common elements that are a part of all microprocessor controllers.

9.4 Field I/O

Microprocessor-based controllers perform control functions only when they receive inputs that can be read by the control program. Inputs received by a controller can be digital, analog, and, in a few cases, of an optical form. A discrete input is a signal received by the controller that is either on or off. A controller will determine if an input signal is received by reading the voltage across the terminals for the discrete input point. If voltage is present, the signal is determined to be on or "high," and "low" if the voltage is off. An analog signal is the output of a field transmitter that is recording a process variable. The signal is received by the controller and it is called, appropriately, an analog input. The analog input can be read in as a 4–20 mA current or it can be read as a 1–5 V dc voltage (these are the most common, but there are other less-common analog input ranges). Before an analog signal can be used, it is converted by the controller to a digital format. The third type of signal that can be employed by a microprocessor controller is the digital signal. The digital signal can represent a range of values or can be a simple on/off reading. The digital signal is a

FIGURE 9-1 A controller with discrete inputs (DI), analog (AI), and discrete output (DO).

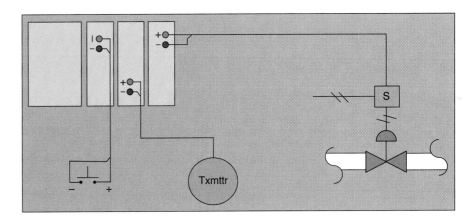

"packet" of information that contains the process variable information as a series of discrete bits.

Inputs must be connected to the controller by field wiring. The field wiring provides the path for the signal to the controller, but the termination point on the controller for each input must be located. All PLCs (programmable logic controllers) have a chassis, slot, and point assignment to configure I/O, so the termination point must be correctly assigned and terminated to ensure the proper input can be read when it is needed. Figure 9–1 shows a single-chassis, three-slot PLC controller. Each slot is configured to perform a different function: digital input, analog input, and digital output. The proper interpretation of documentation will ensure that the correct "type" of input is terminated in its correct location. A field loop sheet can be one source to provide the information needed to terminate I/O to a controller.

Now that the input is received by the controller, the inputs are converted to a usable form, which is merely the letter of the alphabet converted into a series of digital commands. These digital commands, called **bits,** are grouped as **bytes** and, finally, into **words.** When the computer wants information it can read, it is in the form of words. Still, a basic order of events needs to be established before work can be performed; this task is accomplished internally by the CPU.

9.5 Microprocessor Components

The **CPU** (central processing unit) controls all working memory, **program** control, and arithmetic functions. It has a **bus** with which it can send and receive signals and determine the order in which information is exchanged. The CPU receives inputs, performs any calculations needed, and stores the result in **memory.** The control unit in a computer controls the entire operation of memory allocation and logic functions.

The memory of a computer stores information until it is needed. The memory functions only in the sense that it receives information, stores it, and gives it up on demand. The memory performs no calculations, control functions, or logic functions and it is stored in one or several types of storage elements called drums, disks, core, tape, or registers. The register is the only temporary form of a working memory in that it functions only when power is supplied. Working memory, or

FIGURE 9–2 PLC; notice power supply for loop-powered devices as well as PLC power, multiple I/O cards for analog, and discrete I/O.

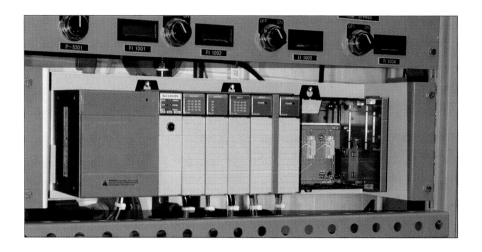

RAM (random access memory), is the fastest way that memory can be temporarily stored and retrieved. ROM (read only memory) is what its name implies—it is memory that can only be read and not written to or changed. RAM is of memory register type, which implies that memory will be lost when power is removed. The other memory components will maintain their memory as long as they are not subjected to altering conditions.

We have covered how the controller will receive its inputs and how they must be terminated to specific locations, or "point assignments" (Figure 9–2). We have read that the inputs, once gathered by an input card for the controller, are converted to a usable format. We have discussed how once inputs are received by a controller, the controller can provide the solution, or outputs, that are needed to accomplish its control function. The order of control actions now depends upon the software that is "running" in the controller.

9.6 Control Algorithm

The control program is often a user-specified software program upon which, in a language the particular controller understands, the "logic" of the program is written. Control languages can vary from the assembly language the computer can directly read to a high-level language (C+, C++, etc.) that is more user-friendly. A compiler is sometimes used to translate a high-level language into machine language. Perhaps the favorite language is the Ladder Logic format, which reminds us of the drawings we have used for years with relay logic. The common thread to all languages is that they must all be read sequentially to be interpreted properly. Sequential operation (sequential logic) forces a controller to perform the control program, in order, from top to bottom within the specified time period. One sequential reading of the program is called a *scan* and the time the program is read in is called the *scan time*.

Remember—a controller cannot provide an output without established inputs and a control program. Once the inputs are read, the program determines the correct response and the outputs are signaled to perform a control action. Outputs can be digital or analog, as inputs are, and the same point assignment procedure and documentation for each still applies. The biggest difference is perhaps that the controller often does not know the response of the field instrument it is "driving." When a controller turns "on" a digital output signal, the solenoid, switch, or annunciation, etc., may or may not respond as it is supposed to. If field technicians know when the field instruments should respond, they will know if the instrument is performing correctly. The same reasoning can be used for an **analog output** signal. The controller does not know how the field device is responding, but the field technician does. Therefore, technicians must be able to read I/O termination points to provide solutions to wiring problems. Advanced field technicians must be able to interpret the ladder language that most PLC controllers will use.

There are several manufacturers of PLCs, but all models function the same way. Regardless of the type used, field technicians will be able to interface with it and understand how it works with their field I/O. A controller can also be a computer that has a visual interface or simply a single-loop controller that acts as a controller for a set number of inputs and outputs. A single-loop controller often has its program stored into permanent memory and it cannot be changed or adapted to fit any working process other than what it was designed for. Although the single-loop controller often has user-selected configurations that can perform a selected routine, it still will have termination points to be connected, it will be configured, and it does not know the response of its outputs. **PID** (proportional, integral, derivative) control actions are often done on a single-loop basis and the single-loop controller does what its name implies: it provides control of the process on a loop-by-loop basis. A single-loop controller may not be as versatile as a software-configured system, but it still can control the process for a process loop with a limited number of I/O.

9.7 DCS Applications

PLCs and computer **terminals** are only two of many forms of controllers used today (Figure 9–3). For larger industrial sites, a **distributed control system (DCS)** will be the primary controller used. Honeywell, Bailey, and Foxboro are a few of the more common DCS systems. For our purposes here, the termination of field instruments will be our only interface with these systems, but if field technicians wish to continue working with instrumentation, they will have to learn working parts of a DCS. A DCS loop contains the same assignments as the loop for a PLC, but the termination point may often be in a "local satellite" building. By using the local termination point, much wiring can be saved because the field wiring will not have to be run back to the control room. These "remote termination points" communicate with the DCS system located in the control room by the use of one or more means. Fiber optics, coaxial cable, and others are only a few ways communication is achieved. The remote locations still have assignments for analog and digital, and inputs and outputs. When DCS systems are used, they will be the primary controllers for the processes at the industrial site. Often there are other controllers that will monitor the DCS operation.

FIGURE 9–3 Control console for boiler house control.

9.8 Emergency Shutdown (ESD) Systems

In areas that have hazardous substances or mechanical operations that pose a danger to people, you may find a PLC acting as a guard over the process. These PLCs are a part of the "critical control systems" often referred to as the emergency shutdown system. The DCS will not have "extra" shutdown actions written into the control program because the setpoints used for controlling are accessible to the operators who can change them. An ESD (emergency shutdown) system is a standalone controller that receives inputs from field instruments, DCS control actions, valve positions, and more, but does not have process variables that can be changed by human interaction. An ESD system monitors the process to observe that a dangerous condition is not existing. If a hazardous situation does arise, the ESD system assumes control of the process and executes an orderly shutdown. Sometimes the field I/O used for the DCS system are the same as those for the ESD system and sometimes the ESD I/O are separate. In extremely critical areas, a field technician will often find multiple sensors recording the same process variable, and all are part of the ESD system. These sensors are evaluated through the use of software to verify accuracy, deviation, and comparisons of like recordings. If a deviation is present, procedures should be available for correction (Figure 9–4).

OSHA is now mandating that critical control I/O should be tested at intervals that will ensure that a failed component of the system will be detected before its failure causes a hazardous result. Dedicated procedures for testing ESD I/O should be available from the site

FIGURE 9–4 On/off control valves; includes limits and solenoids.

where you work. Maintenance of critical I/O should be limited to the standard established for a particular device. With the development of increasingly "smart" I/O, the procedures will surely cover a broader use of communication techniques for monitoring the process and for monitoring the accuracy, repeatability, and all other functions of instrumentation.

9.9 Smart and Peer-to-Peer Control

Field control systems are growing at a pace that will ensure a demand for instrument and control technicians who understand and can evaluate all components of a system. There are "new" devices coming into the market that will expand upon existing knowledge. New I/O will demand that a field technician is fully aware of the fundamentals of instrumentation, knows how a control scheme functions, and is capable of using microprocessor-based communicators to calibrate. A field technician who knows the fundamentals has the greatest potential for an outstanding future in the controls field.

The future of instruments for the process industry will continue to advance along the "smart" guidelines established by industries deciding that digital I/O is the future. The instruments of the future will primarily be smart devices that will be capable of performing any or all of the measuring fundamentals that we have covered. Smart and digital instruments should become a welcome sight to the field technician who understands microprocessor-controlled devices and/or systems.

The future of controllers for the process environment will also continue to advance along the microprocessor lines, but, even now, new devices are beginning to be used for controllers. Some new controllers are not centralized—each device is smart and can initiate communications with other devices. This is the true concept of **distributed control**—the trend of future control systems. Imagine a control system in which each device is capable (smart) of initiating communication, making decisions, giving a response, etc. Such a system, when configured, will have no central controller that governs the process. Each device will govern its own assignment. Such systems will demand accuracy to function properly. Such systems will also demand that the field technicians are equally accurate and precise. The order of accuracy that microprocessors demand ensures that, with the proper knowledge, any problem is solvable. Our job is to ensure that the demands of the microprocessor-controlled systems are met.

This type of peer-to-peer communication and control is being implemented today. Some facilities have decided that peer-to-peer systems will be installed for their noncritical I/O, replacing traditional analog and digital control systems. In effect, this creates a field-distributed "controller." Anyone who performs work on such systems must understand the action of the control system as governed by devices rather than by a centralized controller.

■ SUMMARY

The many applications and types of controllers alone are a field of study. This chapter presented information on the objectives, components, and operation of an automatic controller. Too many field instrument workers see the controller as a source of fault, lacking understanding of its capabilities and limitations. This chapter presented the necessity of providing adequate I/O and establishing the need to determine field I/O accuracy. This chapter could not, in the space limitations of this text, provide in-depth, detailed knowledge into the languages, I/O configuration, settings, or control logic(s) used; such study is an independent requirement.

■ REVIEW QUESTIONS

1. What signal type is normally read in by the controller as a variable signal?

2. What signal type is normally read in by the controller as an on/off signal?

3. What signal type may represent a range of values or may be an on/off-type signal?

4. What is the primary purpose for a centralized distributed control system (DCS)?

chapter 10

Fundamentals of Control

■ OUTLINE

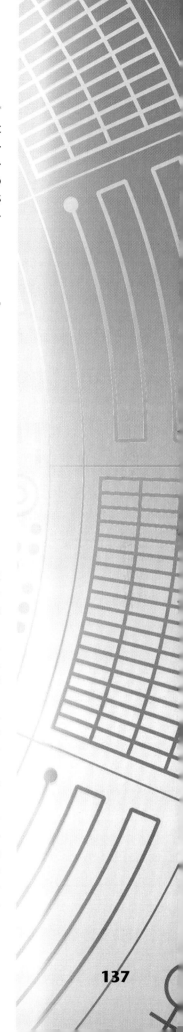

■ OVERVIEW

This chapter provides the methods an automatic controller uses to allow for automatic control. The different forms of control are discussed and explained so that they are readily identified from basic components and functions. Understanding the control-loop designations also provides information as to the devices required to enable the control loop to perform correctly. The required devices that are used in the control loop are the devices studied in the field of instrumentation, and in this text, and therefore must be installed, calibrated, and maintained accurately.

■ OBJECTIVES

After completing this chapter, you should be able to:

- Identify and explain the operation of open- and closed-loop control.
- Explain the effects of resistance, capacity, dead time, and transfer lag.
- Describe the operation of three types of control signals (analog, discrete, and digital).
- Explain the operation of two-position, differential gap, and time cycle control.
- Discuss the operation of throttling control.
- Discuss proportional, proportional-integral, proportional-derivative, and proportional-integral-derivative control.

■ INTRODUCTION

10.1 Fundamentals of Control

An often-used definition for *process* is a function or operation utilized in the treatment of a material. The operation of adding heat to water is a process, as is the operation of removing heat from water through an exchanger. Processes are happening everywhere we go. Your water heater, car, air-conditioner, etc. all use forms of process control.

Process control is the method by which we regulate a particular process. The control action of a thermostat can be used to regulate the motor/compressor operation, which will directly control the temperature at the desired setpoint. In industrial control, the output of a flow transmitter will directly represent the flow volume that it is measuring. For example, if a flow transmitter transmits an output that is equivalent to 5 GPM, and the desired rate of flow is 4 GPM, we know that some control action has to be taken to reduce the flow rate to the desired setpoint. The control action is what we are now going to study.

Most controllers provide several control possibilities and our objective here is to identify and understand how these controllers control a process.

There are several ways to initiate and control processes: mechanically, pneumatically, electronically, and electrically. The control techniques can be either analog or discrete. A third method that will gain predominance as the technology develops is a pure digital control system where analog and/or discrete control can be represented as a series of discrete bits that can be interpreted to represent analog or discrete signals. The important thing is to remember that regardless of the technique(s) used, the basic control fundamentals remain the same.

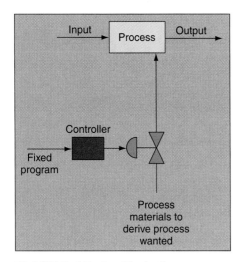

FIGURE 10–1 Block diagram showing a form of open-loop control. The process is measured and a variable is applied to the process but the effects of the added variable are not sampled.

10.2 Control Systems Identified

A control loop is usually an instrument or a group of instruments that are designed, installed, and calibrated to control a process at a desired setpoint. This control loop usually contains a sensor, which senses the process variable; a process that will be controlled; a controller; a setpoint; and a final control element that does the control function.

The response of the system will be dependent on the resistance, capacitance, mass, and dead time that are in the control system. All of these will combine to create transfer **lag,** which is the time it takes for a change to be made to the process. We know that the sensor measures the process variable; the controller then compares it to the setpoint; and the final control element is corrected or adjusted as needed to regulate the process. The control loop functions to control the system. What we need to know is how to make final control device corrections or adjustments so we can predict their response.

There are two basic control loops: **open-loop** control and **closed-loop** control. The open-loop system can be compared to a sensor that measures a process variable and then displays and/or records it. No comparisons or adjustments are made to regulate the process (Figure 10–1). Another control loop must interact to regulate the process—for example, an operator opens a valve manually to lower a tank level when the display indicates a level that is too high. The open-loop control part is the sensor that is displaying a level. The open-loop portion indicates and/or records what is happening with the process, but it causes no corrective action to be taken to regulate the process.

10.3 Closed-Loop Control

The closed-loop control system provides a form of "feedback" to the process under control (Figure 10–2). A process is measured, compared to a setpoint, and a final control element is adjusted accordingly. We now recognize this method of control as the predominant method used today, especially when process control is considered. Closed-loop control uses many, many types of control technologies and control methods that provide feedback and adjustment to a process.

In a process environment that has an established setpoint and control algorithm present in the controller, a change in the process load will determine if corrective action is needed. For example, if the rate of flow in a wastewater treatment facility was increased, the process

FIGURE 10–2 Closed-loop control. The process variable is measured and compared to setpoint and error corrections are adjusted out.

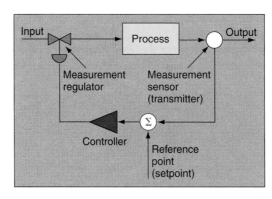

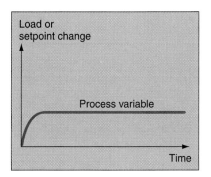

FIGURE 10–3 A low-resistance response curve.

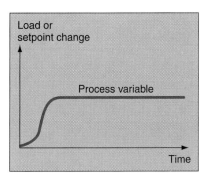

FIGURE 10–4 A typical capacity-response curve.

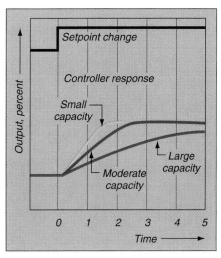

FIGURE 10–5 Transfer lag is also dependent on resistance/capacitance. Notice as resistance and capacitance increase, the transfer lag increases.

would require some adjustment to maintain stability. Therefore, it is easy to see that the change in process loading is the cause of an adjustment to the system. In the past, electrical study of control systems has covered how to set and control the process (i.e., what is the correct setpoint to control the process?). It is more important today to cover the response of a system to a load change and to know how to correctly anticipate how a control loop will measure and respond to a process change and/or a control command. Process loading is the physical change in the process that requires adjustments to be made to control the change in the process.

Resistance is the response characteristic of the system that delays a change in the process variable when a load change is sensed. Any physical material that reduces the rate of energy transfer to a process can be referred to as resistance. Resistance is one element that adds time to the response of a control system (Figure 10–3).

Capacity is the opposite of resistance—It is the ability of a process to withstand change in the process variable. For example, if a flow variable suddenly was indicating a greater flow than wanted, the control response would be to reduce the flow rate. The flow liquid will have a certain inertia that will resist change to a reduced rate. This impetus of flow rate is the capacity of the process variable. This variable will also cause the delay in the change of the process variable to the setpoint (Figure 10–4).

Transfer lag is the time it takes for a process to recover from a change in the process itself. A setpoint change can be viewed as a load change; the characteristics are the same. The time it takes for a process variable to arrive at the setpoint is called the transfer lag (Figure 10–5).

Dead time can occur in all processes, but usually the greater the volume, the greater the chance of seeing dead time. When a change is initiated to a process variable, if there is any time required before a change is observed in the process variable, this is called dead time (not

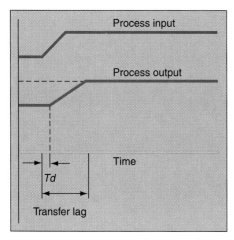

FIGURE 10–6 Dead time *(Td)* and transfer lag shown with respect to input change.

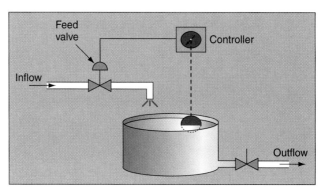

FIGURE 10–7 Self-regulating process.

to be confused with transfer lag). If, in a complicated industrial process, a change is initiated into the system and the process variable remains steady, there is dead time in the system. This usually occurs because the resistance of the system is much greater than the capacity of the process (Figure 10–6).

When the variables of resistance, dead time, and transfer lag are such that a change in the process variable is adjustable and, most important, if the change in the process variable does not exceed the capacity of a system, then **self-regulation** can control the process variable. We have studied how hydrostatic head is used to measure level. Let's use this example to picture a self-regulated process (Figure 10–7).

10.4 Open-Loop Control

Suppose you have a tank that is fed continuously by a water source and you want to maintain a particular level in the tank. The tank has a line flowing from it that cannot be controlled; therefore, the flow out is continuously open. As the tank's level begins to rise, the hydrostatic head pressure increases at the bottom of the tank until the pressure is great enough to force the exit of water, such that it equals the input. The resistance of the line size that restricted the outflow was overcome by the capacity of the system. When the two are equaled, self-regulation is possible (Figure 10–8). The time lag is not present because there is no discernable difference between process load change and process level and there is zero transfer lag because the process is remaining at the desired level. Suppose there is no direct interaction between the input and output of process variables, as the previous example portrays. Then the method of process control and its several methods of control must be utilized.

10.5 Control Signals

Controllers use many techniques to initiate control actions, depending on the degree of accuracy required. Controllers today are available for mechanical, electrical, hydraulic, electronic, and pneumatic methods of control. In each of these methods, the controller's job is to have the system respond to any deviation of the process from setpoint. The ma-

FIGURE 10-8 Five time constants that are determined from the resistance and capacitance of the process.

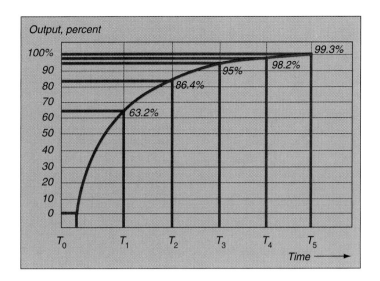

jority of controllers in use today are of the microprocessor variety. This form of controller can accept an analog range as an input, which enables the controller to view the process over a span of designed values, or a discrete input, which allows the controller to see the process at only one point (setpoint). The microprocessor controller uses what is called *digital control* for all of its internal workings. This is a form of converting analog signals to a series of zeros and ones (A/D converters) that represent the same analog value received from the transmitter. This is digital logic, and the controllers of tomorrow will use only digital control logic and digital communication methods. The sensing elements will still see the process over a wide range, but the communication method will not be by an analog signal but rather as digital signals. Digital control is not to be confused with discrete control, which can be referred to as on/off control, which is a form of two-position control.

10.6 Two-Position Control

In two-position control, the final control element is in one of two positions. An example would be a valve that is either fully open or fully closed; there is no throttling of the flow rate to establish control. Two-position control, broadly defined, has three methods of control available: differential gap, time cycle, and on/off control. On/off control is the most important and the most used (Figure 10–9).

The most common method of two-position control is on/off control, which means that when the setpoint is reached by a process variable from either direction, the final control element is adjusted to its opposite position. The thermostat-controlled a/c system is a form of on/off control. As the temperature in a room rises to the setpoint of the thermostat, the thermostat sends a signal to a relay that closes a contact and starts the motor/compressor. The temperature in the room begins to cool after a period of time (dead time) and continues to cool until the setpoint is reached (cooling down-scale; called transfer lag). The actual temperature in the room will fluctuate above and below

FIGURE 10-9 On/off control—
discrete control.

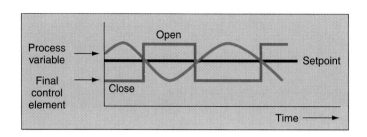

setpoint by a value referred to as *differential gap.* This gap is desirable because it will "dampen" the input signal, which simply means that the cycling of the unit going on and off will be reduced. When on/off control is used, several conditions about the process are known immediately: (1) precise control is not needed; (2) the process must have enough capacity in itself so that the final control element can keep up with process swings; and (3) the process change is relatively small when compared as an energy change to the existing energy of the process.

10.7 Differential Gap Control

Differential gap control is exactly as it sounds: The final control element is adjusted according to a differential gap around the setpoint. Many homes use this method of control to keep the furnace from cycling on and off too often. This results in a wider swing of the process variable (room temperature) than would occur with on/off control but, in this case, the difference is not noticeable.

10.8 Time-Cycle Control

In time-cycle control, a time cycle is established, often by trial and error, that is used to regulate the process variable. An example would be when the temperature setpoint is reached, the final control element is adjusted to its opposite position for a specified time period regardless of the actual change in the process variable. This is an adequate method of control for a system with a significant amount of dead time. If an excessive amount of dead time is present in a system, the process variable will not show any change until the process has absorbed an excessive amount of capacity energy. This condition will cause a delay in response to the control signal as well as an over-desired differential gap. An example is an a/c system that is turned on when the temperature rises to setpoint. The unit will remain running for 1 hour regardless of the temperature inside the room. It is easy to see that the temperature inside the room will be well below the setpoint and a significant amount of time will be needed for the temperature to rise again to setpoint. By trial and error, an adequate time cycle can be established.

Two-position control is by far the easiest and least expensive method of control. Mechanical, pneumatic, electric, and hydraulic devices are easily used in such control methods. For more complicated control actions—those that demand a greater accuracy, such as those used in industrial control—some method of final control is needed. The throttling of final control elements is the answer. This is often referred to as **proportional,** or throttling, **control.**

10.9 Throttling Control

Throttling control occurs when the final control element is positioned somewhere between fully open and fully closed. Its purpose is to throttle the process variable to a rate that can be controlled by the controller.

Proportional control is the usual form of control when the position of the final control element is determined by the relationship between the process variable and the setpoint. Proportional control is used when a smoother method of control is needed than on/off control can provide. The final control element movement is directly proportional with the amount of deviation from the setpoint. Likewise, there is no movement in the final control element unless a change in the process variable is measured. It is easy to follow this form of control, but there is another variable that must be considered (Figure 10–10).

At what point does the final control element reach its fully open or fully closed position? We now know how proportional control applied to a tank level would act when a setpoint for control is set at 50% of span measured by the level transmitter. If the tank level drops below 50%, the final control element is closed some amount to raise the level in the tank. How much is the valve closed? Even more critical, when is the valve signaled to swing fully closed and, likewise, fully open? If the proportional band is set at 20%, the valve would swing fully closed when the setpoint indicated 40%, and would be open fully at 60%. This 20% range of control action is referred to as the **proportional band.** At a range greater than 40% and less than 60%, the valve is at some point between the two fully open and fully closed positions.

When the proportional control method is used by itself, the proportional band must be established to allow efficient control. If the band is set too narrow, the final control element will often act as a on/off controller rather than as a throttling type of controller. If the proportional band is set too high, the controller will not respond to small disturbances in the process variable and the process will not be controlled at setpoint.

Another variable that will come into play with proportional control is gain, or **sensitivity.** Gain is defined as the ratio of output divided by input of a control device and is often explained by using the ratio of two resistances controlling an operational amplifier. Mathematically, gain is the reciprocal of the proportional band that is indicated as a decimal value. Practically, gain is determined by the relationship of change in a process variable to the position of the final control element. In our level-tank example, the proportional band is 20%. Mathematically, the gain is the reciprocal of the proportional band; therefore, the gain would equal 1/0.2 = 5. Perhaps an easier way to calculate gain is to use:

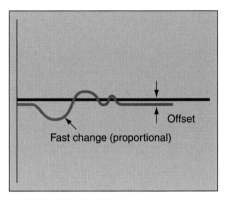

FIGURE 10–10 Proportional control only with resulting offset shown.

FIGURE 10–11 Proportional band with resulting gain shown.

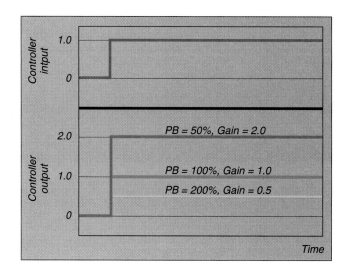

$$\text{Gain} = \frac{\text{Controller output (valve change)}}{\text{Controller input (process variable)}}$$

Once the controller is installed, the gain is set and cannot be changed unless the proportional band is changed (see Figure 10–11). Another way to picture gain is to realize that if a gain equal to 1 is used, as the process variable changes its input measurement value by one unit, the output final control element is changed by one unit. If the gain equals 2, the process variable changing by one unit would cause a final control element response of two units, etc.

A proportional controller may also have a bias variable. Bias is used to allow the output to be set at a predetermined value even if the setpoint is changed. A bias is usually used when the desired output is 50% when the process variable reaches setpoint. A bias usually indicates that the desired valve position is 50% of its opening range given a 50% signal from the controller. We studied this earlier as a linear relationship. The gain is predictable for a proportional only controller and can be figured as follows. Let's use a proportional band of 100% to assume that we want control over the entire range of our process:

$$O/P = [100/\% \text{ proportional band}] \times [\text{setpoint} - \text{measurement}] + [\% \text{ bias}]$$

If the desired level was 50% with a bias of 50%, when the process variable measurement equals the setpoint (0 deviation), the valve position is 50%. When the process variable is measured at 60%, then:

$$O/P = [100/100] \times [-10] + [50] = 40\%$$

This implies that the output of the controller will be 40% of range. Likewise, if the process variable was measured at 40%, then:

$$O/P = [100/100] \times [10] + [50] = 60\%$$

This implies that the output of the controller will be 60% of range.

One fact remains for gain and, therefore, the proportional band when used in proportional control: The higher the gain and, therefore, the lower the proportional band, the more the output will move when subjected to a change in the process variable.

An automatic controller cannot exist without proportional control being implemented. Proportional control tries to return a process variable to its setpoint when a process change is detected. If a process variable change is introduced to a process, the difference is detected between the process variable and the setpoint and corrective action is taken. The controller is always responding to a change. It must measure the process as it is now to initiate a change at some point later; the result is a controller that uses proportional-only control cannot return to its original settings because the bias settings force the output to equal the bias when zero deviation is present.

The difference between process variable and setpoint that is present in a system after the system has stabilized is called the **offset.** The offset in a process can be calculated by:

$$\Delta e \ [\% \ \text{proportional band}/100] \times \Delta pv$$

where Δe is the change in offset and Δpv is the change in the process variable.

It is easy to see that as the proportional band approaches zero, the gain approaches infinity and the offset will approach zero. As we studied earlier, a controller with a high gain acts as an on/off controller. It is easy to see now that a low-proportional band has a high gain and, therefore, a low offset since a controller configured as such will only be able to do on/off control with a minimal amount of dead-time delaying response.

In addition, as the proportional band is increased, the gain is decreased and more offset is introduced into the system. If the proportional band is increased to the full range of the process variable, the gain will be low enough that once the system is destabilized, it will never recover until the bias is adjusted manually to stabilize the process.

When a process cannot stand the offset that is present in proportional-only control and an operator is not available to manually adjust the bias setting, another control function must be employed. Reset, when used in a control algorithm, will integrate any difference between the process variable and the setpoint that is present in the system and not just when the system is stabilized until the difference between the setpoint and the process variable is 0. Reset performs the same function as an operator who manually adjusts bias until zero deviation is reached (Figure 10–12).

Proportional-plus-integral control action automatically adjusts the offset that may be present in a process after a process load change has been introduced to 0% deviation. The introduction of integral control action (reset) has the ability to adjust the position of a final control element when there is an error present in the system at a rate that is proportional to the error. The proper amount of integral control action depends on how fast the process variable can respond to or be

FIGURE 10–12 Proportional band vs. gain comparisons.

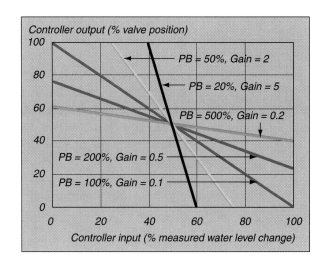

stabilized at setpoint. The integral element in a control system is slow-moving but does eliminate any offset that would be present with proportional-only control.

Proportional-integral-derivative control action is the most effective and efficient control mode. This type of control action will respond to the widest range of control errors. The derivative action produces an output that is proportional to the rate of error change.

The derivative action corrects a final control element by opposing change by an amount that is proportional to the rate of change. Derivative action adds to or subtracts energy from the system. The derivative action dampens the response to a process variable change and provides a stabilizing effect on the system by reducing the oscillations that are present. By reducing system oscillations, the gain can be set at a higher value, which allows a faster response by the controller to a process load change. Derivative action does not correct offset but reduces the amount of overshoot that is present when proportional and integral control are used to stabilize a system after a change has been made.

Being able to identify the methods of control allows us to determine if errors are present and if corrective action is needed. In instrument calibration, we compile a deviation chart that is compared to the ideal case of instrument calibration and then determine if any calibration changes need to be performed. The same calibration procedures can be performed here. The process is monitored and recorded and then compared to a chart to see if any of the control parameters need adjustment.

There is no substitute for practical experience when troubleshooting control systems, but if you understand the methods, you will have a better chance of making an accurate diagnosis. There are several combinations of control system strategies and a detailed discussion is warranted.

10.10 Proportional-Plus-Reset

This type of control is actually a response of two previously discussed control modes: proportional and integral (Figure 10–13). Combining the two modes results in gaining the advantages and compensating for

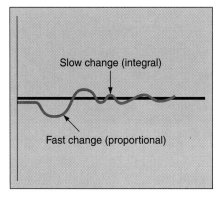

FIGURE 10–13 Proportional-plus-integral (reset) control (PI). Integral control is a time averaging of offset compared to setpoint.

the disadvantages of the two individual modes. The main advantage of the proportional control mode is that an immediate proportional output is produced as soon as an error signal exists at the controller, as shown in Figure 10–14. The proportional controller is considered a fast-acting device. This immediate output change enables the proportional controller to reposition the final control element within a relatively short period of time in response to the error. The main disadvantage of the proportional control mode is that a residual offset error exists between the measured variable and the setpoint for all but one set of system conditions.

The main advantage of the integral control mode is that the controller output continues to reposition the final control element until the error is reduced to zero. This results in the elimination of the residual offset error allowed by the proportional mode. The main disadvantage of the integral mode is that the controller output does not immediately direct the final control element to a new position in response to an error signal. The controller output changes at a defined rate of change, and time is needed for the final control element to be repositioned.

The combination of the two control modes is called the proportional-plus-reset (PI) control mode. It combines the immediate output characteristics of a proportional control mode with the zero residual offset characteristics of the integral mode.

Proportional-plus-reset controllers act to eliminate the offset error found in proportional control by continuing to change the output after the proportional action is completed and by returning the controlled variable to the setpoint. An inherent disadvantage to proportional-plus-reset controllers is the possible adverse effects

FIGURE 10–14 Response of proportional-plus-integral (reset) control.

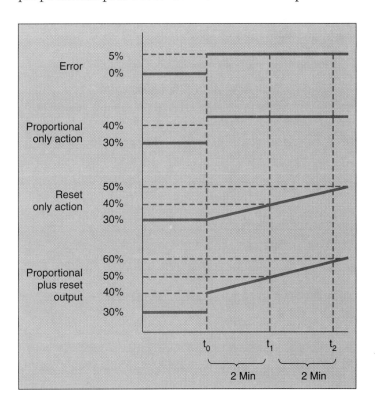

caused by large error signals. The large error can be caused by a large demand deviation or can occur when initially starting up the system. This is a problem because a large, sustained error signal will eventually cause the controller to drive to its limit, and the result is called **reset windup.** Because of reset windup, this control mode is not well-suited to processes that are frequently shut down and started up.

10.11 Proportional-Derivative

Proportional-plus-rate describes a control mode in which a derivative section is added to a proportional controller. This derivative section responds to the rate of change of the error signal, not the amplitude; this derivative action responds to the rate of change the instant it starts. This causes the controller output to be initially larger in direct relation with the error signal rate of change. The higher the error signal rate of change, the sooner the final control element is positioned to the desired value. The added derivative action reduces initial overshoot of the measured variable and, therefore, aids in stabilizing the process sooner. This control mode is called proportional-plus-rate (PD) control because the derivative section responds to the rate of change of the error signal.

Derivative cannot be used alone as a control mode. This is because a steady-state input produces a zero output in a differentiator. If the differentiator was used as a controller, the input signal it would receive is the error signal. As just described, a **steady-state** error signal corresponds to any number of necessary output signals for the positioning of the final control element. Therefore, **derivative action** is combined with proportional action in a manner such that the proportional section output serves as the derivative section input.

Proportional-plus-rate controllers take advantage of both proportional and rate-control modes. As seen in Figure 10–15, proportional action provides an output proportional to the error. If the error is not a step change, but is slowly changing, the proportional action is slow. Rate action, when added, provides quick response to the error.

Proportional-plus-rate control is normally used with large-capacity or slow-responding processes such as temperature control. The leading action of the controller output compensates for the lagging characteristics of large-capacity, slow processes. Rate action is not usually employed with fast-responding processes such as flow control or noisy processes because derivative action responds to any rate of change in the error signal, including the noise. Proportional-plus-rate controllers are useful with processes that are frequently started up and shut down because they are not susceptible to reset windup.

10.12 Proportional-Integral-Derivative

For processes that can operate with continuous cycling, the relatively inexpensive two-position controller is adequate. For processes that cannot tolerate continuous cycling, a proportional controller is often employed. A proportional-plus-reset controller can be used for processes that can tolerate neither continuous cycling nor offset error. A proportional-plus-

FIGURE 10–15 Response of proportional-plus-integral (reset) control.

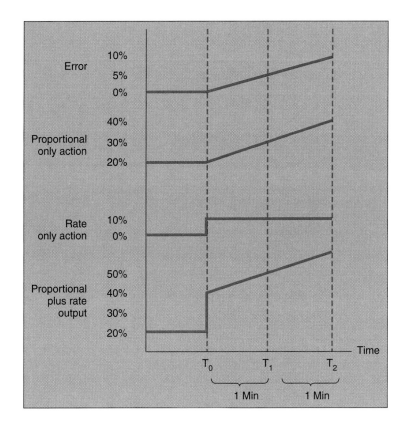

rate controller is employed for processes that need improved stability and can tolerate an offset error.

However, there are some processes that cannot tolerate offset error, yet need good stability. The logical solution is to use a control mode that combines the advantages of proportional, reset, and rate action. When an error is introduced to a PID controller, the controller's response is a combination of the proportional, integral, and derivative actions.

Assume the error is due to a slowly increasing measured variable. As the error increases, the proportional action of the PID controller produces an output that is proportional to the error signal. The reset action of the controller produces an output whose rate of change is determined by the magnitude of the error. In this case, as the error continues to increase at a steady rate, the reset output continues to increase its rate of change. The rate action of the controller produces an output whose magnitude is determined by the rate of change.

As you can see from the combined action curve in Figure 10–15, the output produced responds immediately to the error with a signal that is proportional to the magnitude of the error and that will continue to increase as long as the error continues to increase. You must remember that these response curves are drawn assuming no corrective action is taken by the control system. In actuality, as soon as the output of the controller begins to reposition the final control element, the magnitude of the error should begin to decrease. Eventually, the controller will bring the error to zero and the controlled variable back to the setpoint.

FIGURE 10–16 PID controller response curve.

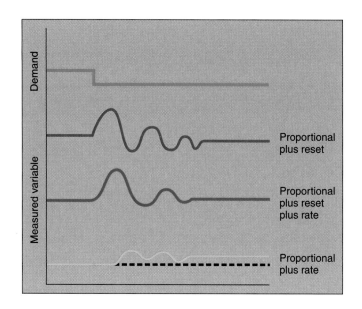

Figure 10–16 demonstrates the combined controller response to a demand disturbance. The proportional action of the controller stabilizes the process. The reset action, combined with the proportional action, causes the measured variable to return to the setpoint. The rate action, combined with the proportional action, reduces the initial overshoot and cyclic period.

▤ SUMMARY

This chapter provided the basic control-loop objectives utilized in automatic control systems. If you can identify the components, devices, and control strategies, you will have a complete understanding of control system operation. Although this chapter does not provide detailed information on the applications of the different forms of control, it does provide suf-

ficient examples of the requirements for each type. All of the devices and physical relationships studied previously interact to create a "system" of operations. This chapter provides one more "tool" to help you understand the control-loop configurations and to help you determine component requirements.

▤ REVIEW QUESTIONS

1. Give a brief description of the devices, variable(s), and instructions that are required for an automatic control loop.

2. What are the two basic types of control loops?

3. What device initially determines if a process loading change requires the need for adjustment to the final control element?

4. What device and resulting function are not included in open-loop control when compared to closed-loop control?

5. When the final control element is a control valve and the position of the control valve is desired to

be somewhere between fully open and fully closed, the resulting control action is referred to as what type of control?

6. If the gain of a controller is 10, what would be the resulting proportional band value?

7. In a controller with a control valve as a final control element, and a proportional band of 20%, at what values above and below setpoint would a final control element be in a fully open or fully closed position?

chapter 11

Fundamentals of Analytical pH Measurement

▪ OUTLINE

■ OVERVIEW

Analytical measurement techniques and equipment are commonplace in industrial process facilities. It is only natural that the set-up, calibration, and maintenance of such devices be performed by the same individuals who work on the existing control system. This chapter provides a brief description of the requirements and functionality of analytical measuring equipment and details pH measurement requirements. pH measurement is a process that requires the same level of accuracy as general instrumentation. The resulting signals can also be utilized in the process control system rather than a standalone application.

■ OBJECTIVES

After completing this chapter, you should be able to:

- Describe the operation of analytical pH measurement.
- Determine if a solution is acidic or alkaline from pH measurements.
- Explain the cleaning of a pH analyzer measurement probes.
- Install, set up, and calibrate a pH analyzer.
- Perform temperature compensation on pH measurements.

■ INTRODUCTION

11.1 Fundamentals of Analytical pH Measurement

Analytical measurement instruments (**analyzers**) are another type of device that gathers process information. An analyzer can be defined as an automatically operating, analytical measuring device that monitors a process for one or more chemical compositions and/or physical properties. Analyzers can be used for control or for monitoring purposes. The measurement process of analyzers is generally more involved than that of the typical devices and process physics we have studied earlier. Generally, analyzers involve some sort of chemical as a process variable or as a means to aid in the measurement of the analyzer (Figure 11–1).

An analyzer provides the methods to improve productivity, reduce environmental impacts of the facility, improve safety, and monitor the process. The past few years have seen an explosion in the types of analytical measurement and analyzers utilized. Unfortunately, there is no single analyzer that performs all measurements needed. The result is a variety of analyzers that we must become familiar with. For our purposes, we will concentrate on pH measurement. Remember, all analytical measurement requires the same basic set-up and maintainance that all other instruments require. The difference is in how we handle and set up the analyzers with chemical and temperature variations rather than our typical process measurement physics.

11.2 pH Measurement

pH is a unit of measurement that determines the degree of acidity or alkalinity of a solution. It is measured with a scale of 0–14. The "p" in the term "pH" is defined as the mathematical equation meaning "negative logarithm." The "H" is the chemical symbol

FIGURE 11–1 MDA scientific
analyzer for phosgene.

for hydrogen. Therefore, pH stands for the negative logarithm of the hydrogen activity and is given by the equation:

$$pH = -\log[H+]$$

The equation provides a method to extract the amount of activity in an acidic or basic form in terms of hydrogen ion activity. If the pH value is less than 7, the solution is acidic; if the pH value is greater than 7, the solution is basic.

A pH measurement analyzer contains four key areas of study: (1) the process sampled; (2) the amplifier; (3) the probes, which are called the *reference;* and (4) sampling electrodes. The process sampled is just what it implies—the process in which a sample is needed to determine pH level(s). The amplifier provides a method to convey the measurement reading into a signal that can be used by the controller or user. The pH electrodes are constructed of a special blown glass that provides the specific resistance requirements that determine the desired outputs of the probes. A pH electrode's glass composition is generally composed of alkali metal ions. The alkali metal ions and the hydrogen ions in the process sample create a exchange reaction between the dissimilar ions, which creates a potential difference. This difference, measured in millivolts, creates the measured variable reading, which is converted to a signal used by a controller.

A **buffer** solution is also employed. The buffer solution has a constant pH value and has the ability to resist changes in pH levels. A buffer solution is used to calibrate a pH measurement system. Most pH meters require periodic calibration at several specific pH levels. One calibration

FIGURE 11–2 pH analyzer station set up for different sample points throughout a local plant.

will be performed at the pH value of 7 (0 mV at 25°C) called the *isopotential point* and another at a pH of 4 or 10. It is important to choose a buffer solution as close as possible to the pH level desired for calibration.

The most common pH electrode is the combination type, which actually has two electrodes encased within one body. The two electrodes are the reference and measuring (sampling) probes. A sampling probe is actually a glass electrode that is technically a battery. This battery generates a voltage that is dependent upon the pH of the solution in which it is immersed. The reference electrode is also a glass probe that performs like a battery, but unlike the sampling probe, the reference electrode voltage does not vary with pH. This battery reference, often called the *rest potential,* is used as a reference to compare with the pH probe's measurement. The comparisons is where the process variable measured is obtained (Figure 11–2)

The reference electrode is the most complicated portion of the pH analyzer. If problems during set-up or measurement occur, they generally are related to the function of the reference electrode. Fouling or contamination is often the culprit for an analyzer's faulty readings. You should take care when you handle an electrode—wear gloves. The glass portion of the electrode should not contact contaminated surfaces, especially your unprotected hands. With the small potential differences measured between the reference and sampling probes, it is easy to see how the contamination of the probes will affect the final readings. The ions created due to contaminents affect readings just as the ions in the measured solution do.

The best way to understand and grasp the function of the reference probe is to think about the voltage-measurement process on a battery. A reading cannot be obtained if only one lead is connected to the battery. If both leads are attached to the battery, a measurement can be

FIGURE 11-3 Glass sample and reference electrodes.

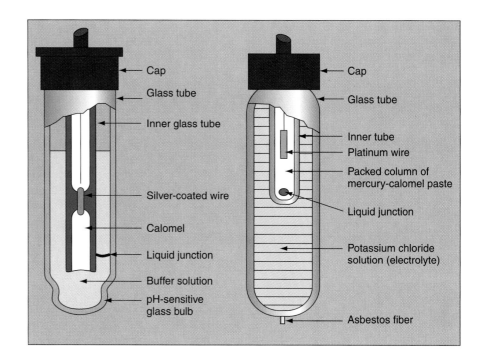

taken. The reference probe is just like the reference lead (ground) lead connected to a battery. When the second lead is attached (sampling probe), the circuit is completed and a measurement can be taken. This measurement is read in millivolts. The millivolt reading is used to determine the pH value of the solution measured as determined by the "+" or "−" ions (Figure 11–3). In practical applications, the reference and sampling probes are submerged in the same solution. The filling solution of the reference probe flows into the sample and completes the circuit. The reference probe must be submerged into the sample to obtain a reading.

11.3 Unbalanced Electrical Charge

An ion is the result of an unbalanced electrical charge in an atom or group of atoms or molecules. Adding electrons creates a negative charge, while removing them creates a positive charge. The unbalanced ions allow us to make a measurement by comparing the difference of the measured variable to the reference probe measurement, which should be stable and known. For example, a solution with a pH level 8 contains ten times more positive hydrogen ions than present in a solution with a pH reading of 7. This holds true throughout the pH level measurement scale. Every unit of change in pH represents a tenfold (1×10^1) change in hydrogen ion concentration.

The pH level is measured on a scale of 0–14. A pH reading of 7 is as a *neutral* solution; pure, distilled water is a neutral solution. The distillation process ensures that the water contains no impurities that will affect the neutral pH reading of the solution. A neutral reading predicts that there is a balance of hydrogen ions and hydroxyl ions in the solution. Solutions with a pH level less than 7 are said to be more *acidic,* which implies that they contain more hydrogen ions. Solutions

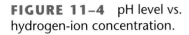

FIGURE 11–4 pH level vs. hydrogen-ion concentration.

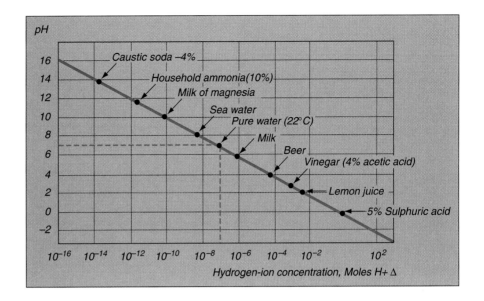

with pH values greater than 7 are *alkaline,* which means they contain more hydroxyl ions (Figure 11–4).

A neutral solution can be referred to as a buffer solution. A buffer solution is used as a reference reading to determine the pH level of the measured solution. The resulting difference, a measured value between the measuring probe and the reference probe of the buffer solution, can be compared to the reading of the measured solution to obtain a pH level. As with all calibrations, there must be a standard to which the measured variable can be compared. Some of the solutions used for comparison have pH values of 4, 7, or 10: one neutral solution, pH = 7; one acidic, pH = 4; and one alkaline, pH = 10.

11.4 pH Measurement Equipment Set-Up and Calibration

To install a pH analyzer, the electrodes must first be thoroughly cleaned with distilled water to remove any contaminates. Electrodes should be rinsed between samples with distilled or de-ionized water. Never wipe an electrode to remove anything from it—rinse the probe to clean and decontaminate. The level of fluid in refilling-type electrodes should always be kept two-thirds full. Always check and use the probes in the vertical position.

Once clean, rinse the electrodes with the buffer solution with a pH = 4. Then insert the probes into the buffer solution (pH = 4) and record the readings of the analyzer. If necessary, make adjustments to the analyzer's readings to match those of the buffer solution. Then rinse the probes once again with a neutral solution (pH = 7) and repeat the process. Repeat the steps with a solution of pH = 10 and record/adjust if necessary. At least a two-point check is required; most facilities require a three-point check. After you have taken the readings, check sensitivity. A sluggish reading is an indication that

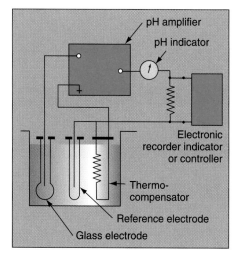

FIGURE 11–5 pH measurement circuit.

Table 11–1 Temperature and pH Temperature-Influenced Compensation

	Temperature	
pH	Above 25°C	Below 25°C
pH	Above 25°C	Below 25°C
Above 7	Subtract	Add
Below 7	Add	Subtract

the measurement probe needs to be replaced. To complicate matters, a temperature-compensating element may also be required to compensate for pH differences due to effects of temperature changes (Figure 11–5).

The millivolt output of pH electrodes varies with temperature. The magnitude of the variation is directly dependent on the temperature change and the pH value of the solution being measured. At a pH reading of 7 and a temperature of 25°C, there is zero temperature compensation needed. At any other reading, temperature compensation is adjusted by 0.03 pH error/pH unit/10°C (Table 11–1).

The actual voltage measurements are temperature-compensated by applying the values in Figure 11–6 and using the directional compensations shown in Table 11–1. Figure 11–7 displays the linear relationships.

FIGURE 11–6 pH vs. temperature corrective adjustments in millivolts.

		pH Value										
		2	3	4	5	6	7	8	9	10	11	12
Temperature	5	.30	.24	.18	.12	.06	0	.06	.12	.18	.24	.30
	15	.15	.12	.09	.06	.03	0	.03	.06	.09	.12	.15
	25	0	0	0	0	0	0	0	0	0	0	0
	35	.15	.12	.09	.06	.03	0	.03	.06	.09	.12	.15
	45	.30	.24	.18	.12	.06	0	.06	.12	.18	.24	.30
	55	.45	.36	.27	.18	.09	0	.09	.18	.27	.36	.45
	65	.60	.48	.36	.24	.12	0	.12	.24	.36	.48	.60
	75	.75	.60	.45	.30	.15	0	.15	.30	.45	.60	.75
	85	.90	.72	.54	.36	.18	0	.18	.36	.54	.72	.90

FIGURE 11–7 Millivolt variations given temperature change and pH value.

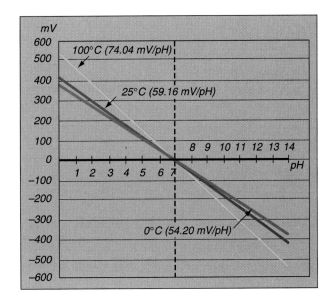

SUMMARY

This chapter presented the basic concepts involved with analytical measurement and a more detailed explanation of the pH measurement process. pH measurement does bring into play several new variables that are not associated with instrumentation, but this chapter presented the set-up requirements, which do demonstrate an association.

REVIEW QUESTIONS

1. Analyzers generally perform what type of measurement (Hint: not direct or indirect measurements)?

2. What four areas of concern must be addressed if you are working with pH measurement equipment?

3. How should the probes of an analyzer be handled? Why?

4. What solution is used as a neutral solution in pH measurement?

5. A sample returns a result of 11 on the pH scale. Is the solution acidic or basic?

6. What is the purpose of a reference (buffer) solution when setting up pH measurement analyzers?

chapter 12

Fundamentals of Smart Instrument Communicators

■ OUTLINE

■ OVERVIEW

This chapter introduces a smart communicator into the calibration process. A smart communicator is used during the calibration of a smart device and, therefore, is needed to verify its operation. This chapter does not provide a detailed procedure into the calibration process, but rather explains the functions, components, and contributions made to the process. The smart device calibration procedure is a detailed process, but to accomplish the goal of calibration, the working parameters of the smart communicator must be understood first.

■ OBJECTIVES

After completing this chapter, you should be able to:

■ Explain the objective(s) for using a smart communicator.
■ Describe how a smart communicator is used to observe calibrated ranges.
■ Explain how data are retrieved and entered using a smart communicator.
■ List the features of a smart communicator.
■ Provide a brief description of the communicator's role in the calibration procedure.

■ INTRODUCTION

12.1 Fundamentals of Smart Instrument Communicators

A smart communicator is used to provide a working interface to a smart device. A smart communicator is the "tool" used to adjust zero and span, which were adjusted earlier by turning screws. The smart device to which a communicator is connected, typically a transmitter, is used to perform the function of reading a process variable, like its conventional counterpart. But the smart device is microprocessor-driven and controlled. Communication signals are electrical impulses—currents—so all safety concerns need to be addressed for hazardous areas (Figures 12–1 and 12–2).

A smart device is installed by the same process (impulse) tubing and wiring arrangements used for a conventional device. Being able to "communicate" with a smart device is what separates it from its conventional counterpart. A smart instrument communicator interacts with the smart device, and the **protocol,** or language, is determined by the manufacturer. Some smart devices communicate digitally; others use field wiring to communicate via a carrier signal (still digital communications). The thing to remember is that all smart communicators perform the same working interface function—they allow configuration of a smart device. The "keystrokes" may be different, or of a different sequence, but the same objective exists. A smart communicator should be accompanied by documentation to familiarize the user to its specific keystroke functions. In time, a user will become familiar with the various "menu trees" used to navigate through the different functions for calibrating and checking a device.

12.2 Application

The device we will study here, the Rosemount HART (Highway Addressable Remote Transmission) communicator, is a handheld interface that provides a common communication link to all HART-compatible, microprocessor-based instruments.

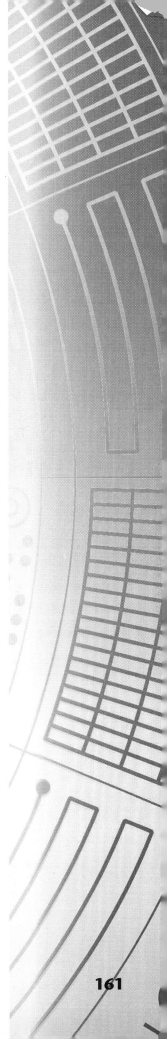

⚠ WARNING

Explosions can result in death or serious injury. Before connecting the HART Communicator in an explosive atmosphere, make sure the instruments in the loop are installed in accordance with intrinsically safe or nonincendive field wiring practices.

FIGURE 12-1   Safety awareness.

⚠ WARNING

Explosions can result in death or serious injury. Do not make connections to the serial port or NiCad recharger jack in an explosive atmosphere.

FIGURE 12-2   Safety awareness.

The HART communicator can interface to a smart Rosemount device at the site of the instrument, a junction box or termination point, or in the control room. In short, a HART communicator can establish communications with a smart device when attached anywhere in the signal loop provided there is a minimum of 250 ohms present in the loop. Loop connectors located on the rear of the communicator allow for access to the loop wiring by attaching the communications to the loop wires across polarity. The communication leads are not polarity sensitive. A PC (personal computer) connection can be made through a serial port, also located on the rear of the communicator. The Hart communicator does not measure loop current directly but does use the loop resistance to calculate the loop current (Figure 12–3).

Users can observe the actions and/or functions of the smart device through the liquid crystal display (LCD). When connected to a device, the LCD can display the model number, the instrument Tag Number, and messages. The bottom of the display is reserved for the software function keys (F1–F4), which allow a user to flow through the functions of communicating to a smart device. "Help" functions are available for certain routines and Help is displayed on the "F" keys when available. Function keys change as the user goes through the different menus that are displayed. Depending on the menu, the function keys will display a set of instructions to execute a command, exit the menu, go to the home menu, etc. Function keys are not the only method of selecting information and/or commands.

Alphanumeric and shift keys perform a fast selection of menu items and data entry. To enter data, the shift keys are used in conjunction with the letter keys and the strings of letters are entered separately (Figure 12–4).

The HART communicator, when turned on, will generally display one of two menus: one if a device was found when the communicator was turned on, and another if a device is not found. If no device is found, the communicator should say so and will display the menu for off-line configuration.

If a device is connected, the menu for on-line should be displayed along with the relative device information. If a device is not found, you

FIGURE 12-3   Rosemount smart communicator connection ports.

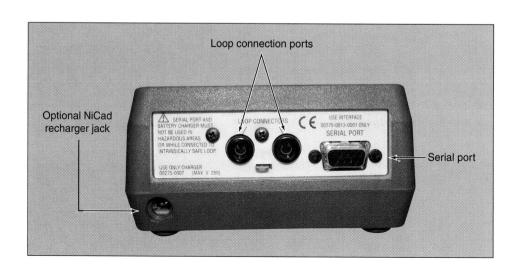

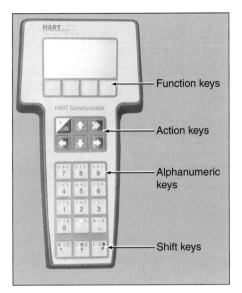

FIGURE 12-4 Alphanumeric keys for data entry.

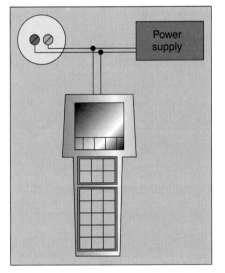

FIGURE 12-5 Loop communications.

FIGURE 12-6 Bench set-up for smart device calibration.

should troubleshoot the connections and/or wiring until you locate the problem. It is not too soon to verify that you are trying to communicate to a "smart" device and not its "dumb" predecessor (Figure 12–5).

Sometimes a specific HART device is not found when connected and the wiring and/or connections are proved to be satisfactory. In these cases, the user is limited to perform data exchange to the device because the device is not stored in the memory module of the calibrator. Generic descriptions will have to be used. For example, if you have an older communicator and a newer device, the communicator will not likely know that such a device exists and will not be able to recognize it. The closest model to the device will have to be chosen for data exchange or a generic menu will be available. The device will accept the information because it should be downwardly compatible, but it may not have the same parameters as the older model (URV, LRV, sensor type, etc.). Never attempt to calibrate a newer device than the calibrator recognizes until you receive proper authorization and acknowledgment from whomever is directing the calibration procedures (Figure 12–6).

A handy feature of the HART communicator is that it allows the storage of multiple devices that are configured off-line and downloads them later. These devices must be configured off-line and stored in the communicator for later use. The manufacturer's devices are listed and, when selected, the model numbers appear. When the field device revision menu appears, the menus continue onward until complete and a new device is selected.

Regardless of whether a device is configured on-line or off-line, the device parameters have to be entered correctly. The process range limits, analog output ranges, and damping selection (if any) must all be entered as specified. The actions of entering data are menu-driven and follow a predefined sequence of steps. Menu trees that enable a user to "see ahead" are available, allowing users to see their selection before it is chosen (Figures 12–7a and 12–7b). If users follow a menu tree they will have few problems with a device being configured incorrectly.

Several menus are available for advanced diagnostics and troubleshooting techniques. These menus should not be activated unless

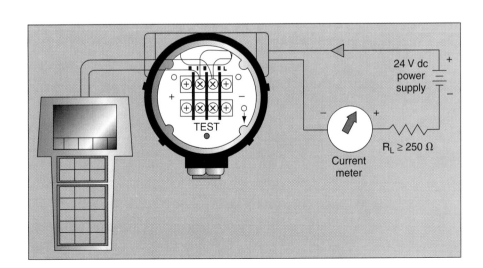

FIGURE 12–7a
Rosemount Model
1151 menu tree.

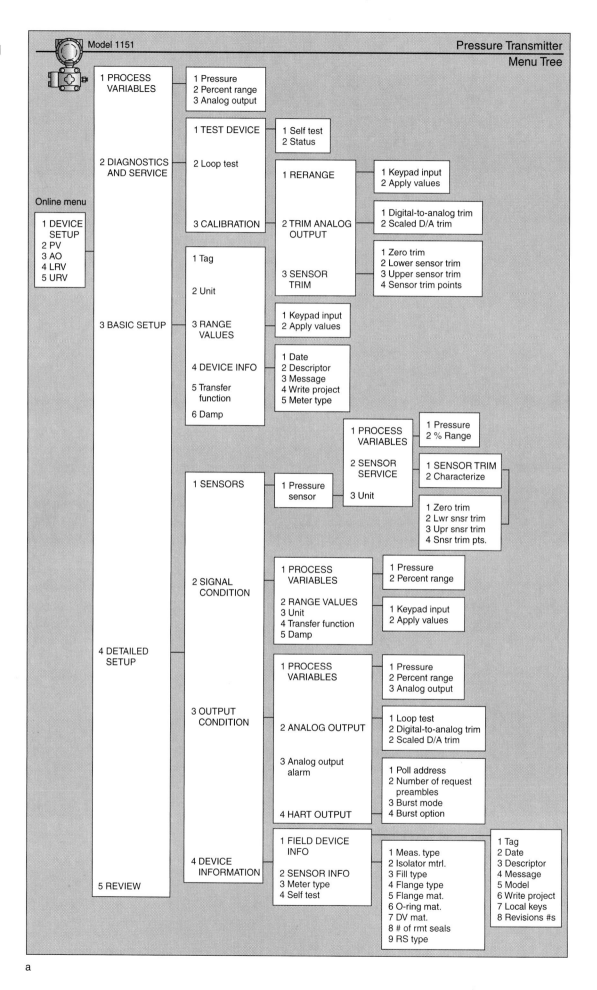

a

FIGURE 12–7b Fast key
sequence.

Function/Variable	Fast-Key Sequence
Analog Output	3
Analog Output Alarm	1, 4, 3, 3
Burst Mode Control	1, 4, 3, 4, 3
Burst Operation	1, 4, 3, 4, 4
Calibration	1, 2, 3
Characterize	1, 4, 1, 1, 2, 2
Damping	1, 3, 6
Date	1, 3, 4, 1
Descriptor	1, 3, 4, 2
D/A Trim (4–20 mA Output)	1, 2, 3, 2, 1
Field Device Info	1, 4, 4, 1
Full Trim	1, 2, 3, 3
Keypad Input	1, 2, 3, 1, 1
Loop Test	1, 2, 2
Lower Range Value	4, 1
Lower Sensor Trim	1, 2, 3, 3, 2
Message	1, 3, 4, 3
Meter Type	1, 3, 4, 5
Number of Requested Preambles	1, 4, 3, 4, 2
Percent Range	1, 1, 2
Poll Address	1, 4, 3, 4, 1
Pressure	2
Range Values	1, 3, 3
Rerange	1, 2, 3, 1
Scaled D/A Trim (4–20 mA Output)	1, 2, 3, 2, 2
Self Test (Transmitter)	1, 2, 1, 1
Sensor Info	1, 4, 4, 2
Sensor Trim Points	1, 2, 3, 3, 4
Status	1, 2, 1, 2
Tag	1, 3, 1
Transfer Function (Setting Output Type)	1, 3, 5
Transmitter Security (Write Protect)	1, 3, 4, 4
Trim Analog Output	1, 2, 3, 2
Units (Process Variable)	1, 3, 2
Upper Range Value	5, 2
Upper Sensor Trim	1, 2, 3, 3, 3
Zero Trim	1, 2, 3, 3, 1

b

the user clearly understands the consequences of selecting a configured device for diagnostics.

Users do not need to verify the calibration of smart communicators, as is necessary with other calibration equipment. The communicator does not contain measuring circuitry and does not measure analog or discrete values directly. Smart communicators are just what their name says—a communicator. A smart communicator is a communications device that provides an interface to a smart device and is not subject to calibration methods. When the communicator is displaying milliamp values, process variables, etc., it is not measuring the variables; it is merely displaying the values that the smart device "tells" it.

The modular construction of most communicators makes it easy to repair and replace the various components. The battery pack, the memory module, and data packs are some of the modules that can be exchanged and/or replaced. Remember to store and exchange the data saved in the memory module and data packs if you need it later for retrieval.

The communicator provided by Rosemount is typical of most communicators since it will accept the alkaline or nicad rechargeable modules as its power supply. A low-battery icon will appear after approximately 1 hour of usage. The communicator should be serviced when this icon appears. Continuing data entry and/or storage under the low-power conditions may result in a loss or corruption of data. Figures 12–8a and 12–8b show the common tasks, functions, and descriptions of the most common devices and procedures performed by the Rosemount communicator.

It is recommended that the loop be placed in manual before any communications are attempted over a field-installed device. This means that operations must place the loop into manual control from an operation's standpoint. Placing the loop in manual allows for communication to a device without causing process swings or deviation. With some communicators, it is critical to do this step first. Merely attaching the communicator to an active process loop will cause the process variable measured by the smart device to "swing" uncontrollably. A prompt appears on most communicators to place the loop in manual—this is a prompt only. By selecting "OK" when the prompt appears, you are not placing the loop in manual, you are acknowledging that you have asked operations to place the loop into manual (Figure 12–9).

12.3 Using the Communicator to Calibrate a Smart Instrument

It is always important to recall that the calibration procedure requires the utilization of certified calibration equipment. The communicator does not have to be certified because it is a communication device used to configure a smart instrument. Therefore, on calibration procedures, the communicator should not be listed as calibration equipment. A detailed study of the calibration procedure using a smart communicator is covered in Chapter 13, but a brief description follows.

To calibrate a smart device, a communicator is used to change the working parameters of the device. The certified test equipment is used to verify that the device is calibrated and performing as required. An input test standard must still be used to supply the device's required input range, and an output standard is used to verify the output signal of the device. The communicator is used to determine if the device is performing the way the test equipment determines.

The communicator is used to apply the working range of the device first as the upper and lower range values in the same working units that the transmitter observes and is calibrated in. Then the communicator is used to verify that the transmitter "sees" the same input supplied to the device as the certified test equipment documents. Last, the communicator is used to verify that the output of the device is performing the same as the certified test equipment displays.

FIGURE 12–8a
Rosemount Model 3051C menu tree.

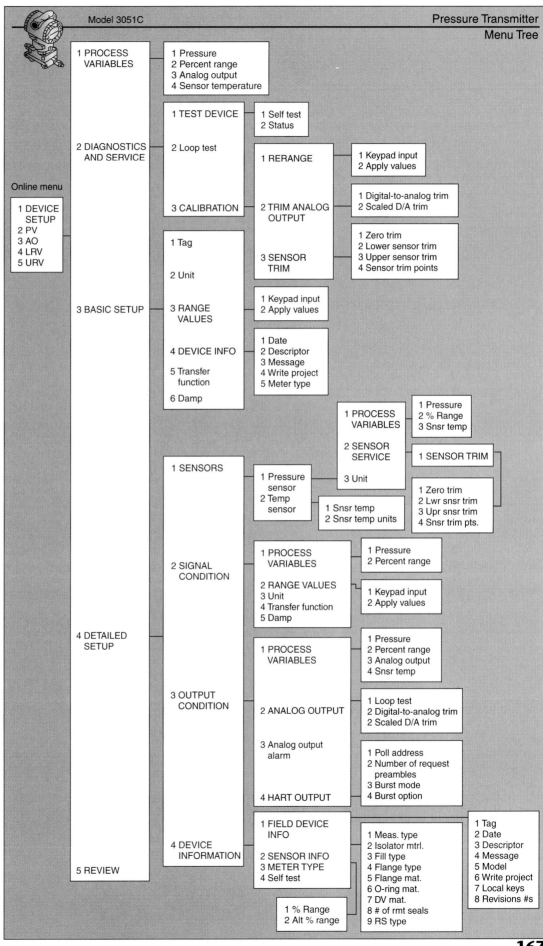

FIGURE 12–8b Fast key
sequence.

Function/Variable	Fast-Key Sequence
Analog Output	3
Analog Output Alarm	1, 4, 3, 3
Burst Mode Control	1, 4, 3, 4, 3
Burst Operation	1, 4, 3, 4, 4
Calibration	1, 2, 3
Clone Data	Left Arrow, 3
Damping	1, 3, 6
Date	1, 3, 4, 1
Descriptor	1, 3, 4, 2
D/A Trim (4–20 mA Output)	1, 2, 3, 2, 1
Disable Local Span/Zero Adjustment	1, 4, 4, 1, 7
Field Device Info	1, 4, 4, 1
Full Trim	1, 2, 3, 3
Keypad Input	1, 2, 3, 1, 1
Loop Test	1, 2, 2
Lower Range Value	4, 1
Lower Sensor Trim	1, 2, 3, 3, 2
Message	1, 3, 4, 3
Meter Type	1, 3, 4, 5
Number of Requested Preambles	1, 4, 3, 4, 2
Percent Range	1, 1, 2
Poll Address	Left Arrow, 5, 1
Pressure	2
Range Values	1, 3, 3
Rerange	1, 2, 3, 1
Scaled D/A Trim (4–20 mA Output)	1, 2, 3, 2, 2
Self Test (Transmitter)	1, 2, 1, 1
Sensor Info	1, 4, 4, 2
Sensor Temperature	1, 1, 4
Sensor Temperature Units	1, 4, 1, 2, 2
Sensor Trim Points	1, 2, 3, 3, 4
Status	1, 2, 1, 2
Tag	1, 3, 1
Transfer Function (Setting Output Type)	1, 3, 5
Transmitter Security (Write Protect)	1, 3, 4, 4
Trim Analog Output	1, 2, 3, 2
Units (Process Variable)	1, 3, 2
Upper Range Value	5, 2
Upper Sensor Trim	1, 2, 3, 3, 3
Zero Trim	1, 2, 3, 3, 1

b

FIGURE 12–9 Field communications and loop wiring paths.

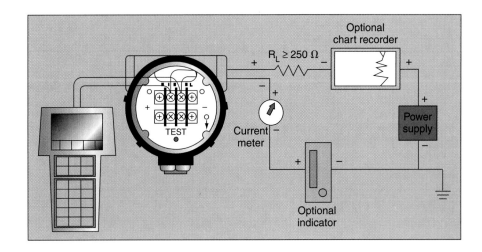

▪ SUMMARY

The use of smart communicators as a calibration aid is commonplace in the industrial environment. Too often, calibration procedures are not performed correctly because the user lacks understanding of the calibration procedure and the communicator. This chapter provided information on the communicator to assist in the calibration process by describing the functions, components, and objectives of the communicator. Knowing how to enter, retrieve, and interpret the data using a smart communicator is the key link in determining if a device is calibrated correctly and performing correctly.

▪ REVIEW QUESTIONS

1. Does a smart communicator have to be calibrated? Explain your answer.

2. Is a smart communicator listed on the calibration data records and procedures as required calibration equipment?

3. How are data entered using a smart communicator?

4. Give the three basic calibration checks performed when using a smart communicator.

5. Using Figure 12–8, give the key sequence used to reach the calibration menu for the Rosemount 3051C pressure transmitter.

chapter 13

Fundamentals of Smart Instrument Calibration

■ OUTLINE

■ OVERVIEW

This chapter provides a detailed look into the calibration requirements for smart devices. The required procedures are given along with supplemental information explaining the purpose for the procedure. The need for a five-point check with a smart instrument is given as a way to verify calibration, but this chapter also shows how to check and verify the calibration range, sensor configuration and accuracy, and output type desired.

■ OBJECTIVES

After completing this chapter, you should be able to:

- List the three steps required to calibrate a smart instrument.
- Describe the function of the select switches found in smart instruments.
- Perform the analog output configuration of a smart device.
- Determine sensor accuracy and perform sensor trim procedures.
- Perform a digital-to-analog trim procedure.
- Identify a multi-drop configuration for digital communication devices.

■ INTRODUCTION

13.1 Fundamentals of Smart Instrument Calibration

A smart instrument is often thought of as an instrument that is microprocessor-based. Many correctly assume this but few understand how this instrument differs from conventional instrumentation. If an instrument is described as being smart, then it does operate differently from conventional analog transmitters. Popular beliefs today indicate that a vast majority of installers and calibrators think there is an "unknown," or "magic" sequence of operation for a smart transmitter. This assumption could not be further from the truth.

We will learn that the construction, installation, and calibration of a typical smart instrument is, in fact, a simpler job than performing the same functions on a conventional analog transmitter. The predominant conventional analog transmitter is the differential pressure type; the same holds true for smart instruments. The examples of step-by-step procedures described in this chapter refer to the smart Rosemount Model 1151 DP or 3051 pressure transmitters.

Communication to a smart instrument is initiated with a communicator. The type of communication device used depends on the manufacturer, and communicators cannot be alternated between different manufacturers. A typical "bench" calibration hook-up is shown in Figure 13–1.

13.2 Bench Calibration Set-Up Example

When you bench-calibrate a smart instrument, you can connect the positive lead of the ammeter to the positive test point and the negative lead of the ammeter to the negative test point. Smart transmitters are often calibrated on the "bench" first, which ensures that the transmitter is in good working order. The hardware switches can also be set so

FIGURE 13–1   Bench set-up for communicating and calibrating a smart device.

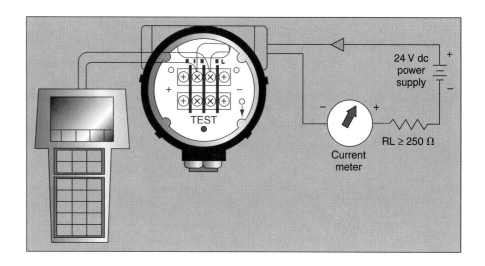

FIGURE 13–2   Switch functions and default positions.

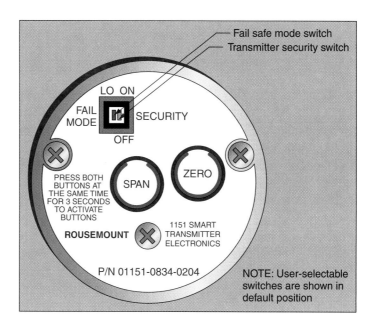

the internal workings of the transmitter will not be exposed to a hazardous atmosphere.

Most smart devices contain switches that allow the persons who are configuring the transmitter to select whether they would like the transmitter to fail "hi" or fail "low." If a device is selected to fail "hi," when the internal diagnostics of the transmitter detect an unrecoverable error, it will usually drive the analog signal to a value greater than 20 mA. Likewise, if fail "low" is selected, the transmitter will output a signal less than 4 mA. These switches are located inside the housing cover and are accessible when the cover is removed (Figure 13–2). These switches are often overlooked when installed.

Smart devices arrive with the switches in their default position. The installer should verify that the fail position will not drive the output signal to a level that is not detectable by monitoring equipment. For example, if a transmitter that is measuring level was set to fail "low" and the alarm points for the level monitored were for high lev-

FIGURE 13-3 Jumper installation for Model 3051.

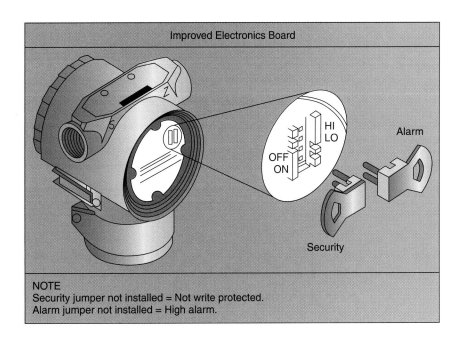

Improved Electronics Board

HI
LO
OFF
ON
Alarm
Security

NOTE
Security jumper not installed = Not write protected.
Alarm jumper not installed = High alarm.

els only, it is easy to see that a transmitter failure would not alarm for a high-level condition. Most DCS (distributed control systems) monitors and/or controllers are capable of detecting a transmitter "over range" error. This would imply that a high or low setting would be detectable as a "bad point" error, but care should be taken to verify the switches are still placed in the proper position.

The security switch is another setting that can be used. The security switch setting requires that it is in the "off" position in order to change, enter, or transmitt configuration data. Once the configuration switches are in their proper position, transmitter configuration/testing can proceed (Figure 13–3).

When the transmitter is connected as a bench calibration or as a field connection, a communicator can be connected. With a Rosemount device, a HART-based communicator would be turned on by pressing the On/Off key. The communicator will search for a device and show that it is connected to a smart device. If it does not, you should troubleshoot the loop to see where the loss of communication is occurring. When connected, there are several recommended test functions that you should perform. Testing is often done when poor loop performance is detected and when a device failure occurs (Figure 13–4).

A communicator test is performed to ensure that the communicator is working correctly. Most communicators perform a self-test when turned on. If an error or failure is detected, the communicator will list a diagnostic message. The important thing to note here is if a communicator is in proper condition, transmitter configuration may be continued.

You should also perform a transmitter test. Although smart devices are continuously self-diagnosing signals and internal diagnostics, a more extensive self-testing routine should be initiated with the transmitter test routine. The transmitter test routine will identify any errors and display messages that will indicate the source of the problem.

FIGURE 13–4 On-line menu that is displayed when connected to a Rosemount Model 1151.

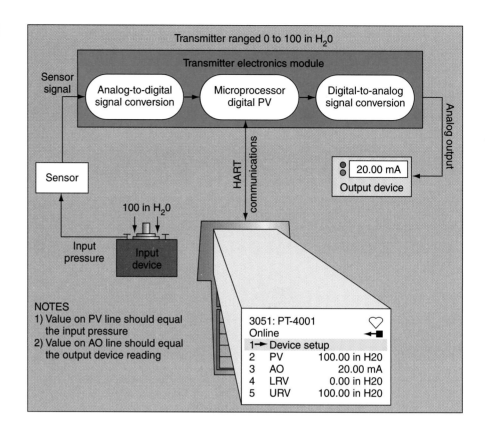

A loop test can be performed to verify output of the transmitter, recorders in the loop, and to ensure the proper wiring of the loop. A loop test should be performed when the device is installed to ensure the proper connections in the loop have been made. Messages will usually be displayed to remind you to put the loop in manual. This function ensures that there are no devices that are being controlled from the transmitter. If a device was being controlled from the transmitter and a loop test was performed, the final control element would swing to its upper and lower range limits. It is easy to see that a working process cannot be subjected to such an extreme variation.

Communicators will allow the transmitter in the loop test mode to output a chosen signal regardless of the input from the sensor. If 4 mA is chosen, all devices in the loop should indicate a signal of 4 mA. If 20 mA is chosen, all devices in the loop should indicate a signal of 20 mA. Sometimes, as with the Rosemount 1151 DP, the output signal will not register 4 mA when it should and the digital trim should be adjusted (to be covered later). When checking the output of the transmitter, the "engineering units" should be indicated to correspond to the working process. Setting the engineering units to match the process allows the calibrator to view the process in working parameters. Therefore, if the engineering units indicated at 4 mA do not correspond to a actual loop indication of 4 mA, a trim is again required.

These steps should be repeated whenever the testing and/or troubleshooting of a device or loop is called for. Whenever you are preparing to send or request data that will disrupt the control loop or change the output of the transmitter, you should place the loop in manual.

Placing the loop in manual implies exactly what it is—the transmitter will no longer provide a basis for loop operation. The prompt that appears in the HART communicator display is a reminder only; placing the loop in manual is a separate function. Once all of these steps are verified and completed, the transmitter calibration can be verified.

13.3 Calibration Procedure

Calibration of a smart transmitter requires some basic steps that are the same for all smart devices. Calibration of the Rosemount 1151 requires three major steps. With other devices and other manufacturers, the following steps should be relatively the same but device calibration procedures should be referenced.

1. Configure the analog output parameters. Do this by setting the upper and lower range limits of the device.
2. Verify that the sensor is accurate during the calibration procedure.
3. Verify that the 4–20-mA output is accurate during the calibration procedure.

13.4 Configuring the Analog Output Parameters Requires Four Basic Steps

Configuring the analog output parameters requires four steps:

1. Set the process variable units of measurement. This allows a calibrator to be able to view the calibration range in the same parameters as the process monitored. Often enough, the smart transmitter is configured from the factory in engineering units other than the process parameters that will be observed during calibration. Several options are available (14 with the Rosemount 1151 DP) and the calibrator is able to select the options from the menu provided. For example, if your calibration procedure calls for you to calibrate the device in "H_2O, you would want to set the device to display measured input values in "H_2O.
2. Re-range the 4 and 20 mA outputs to match the upper and lower range values. This is one of the most common calibration procedures for any transmitter. The smart transmitter is no different. Re-ranging a transmitter's signal allows the transmitter to accurately match the process measured. With the smart transmitter, as with a conventional transmitter, the working range selected by the user cannot be outside factory-allowable ranges. You can check this by making sure the upper (URV) and lower range values (LRV) are "inside of" the upper set limit (USL) and lower set limit (LSL). To re-range the device, enter the desired upper and lower range values using the communicator and then send the information to the calibrating device. The device is now configured to signal its measured variable from the "zero," 4 mA setting up to the URV, and 20 mA. We will discuss the procedure to verify accuracy later.
3. Set the desired output type. This is a user-selectable option that needs attention. We know that not all output signals will be proportional or

linear. For example, some signals such as level will use a linear representation of 50% (12 mA) for a 50% level. We know that flow rates, due to the square-root relationship we studied earlier, are not necessarily a linear function signal; exactly 8 mA = 50%. With a conventional transmitter, there must be a device or method that will convert the signal to a usable value. With a smart transmitter, the user can select a square-root output option that is proportional to flow when activated or a linear option that provides % input = % output.

4. Set damping. A smart transmitter damping allows it to filter out noise and unwanted process fluctuations. An example would be to use electronic damping to "smooth out" the output when there are extremely rapid input variations. High damping values filter out process noise and unwanted fluctuations while low damping values increase the response time of the instrument but can introduce rapid fluctuations of input values. Damping values are adjustable with a smart calibrator from 0–16 seconds in 0.1-second increments (Rosemount 1151).

Steps 1–4 can be performed before process input is applied to the transmitter and before a measuring standard is used to record the device's output. To perform the remaining steps, applied input standards and output standards must be available to supply and record process variables and signals. A standard five-point check is still utilized to identify if any instrument errors are present.

13.5 Calibrating the Sensor

To verify sensor accuracy, an input standard reading is applied and compared to the device's measured variable. For example, if 120 ″H_2O is supplied by an input standard, the 1151 should record that it is reading 120 ″H_2O for a process variable (PV). This is not always the case. Sometimes when a standard is applied to the device, the device will record and display a different value. For example, if 120 ″H_2O was applied and the input standard indicated 120 ″H_2O but the 1151 records the value as 118 ″H_2O, this is a sensor error. You can identify sensor errors by performing the five-point check and verifying that the input standard and the device record identical values. If they do not match, the sensor needs to be calibrated.

To properly comprehend the digital trim function, you must understand that the smart transmitter does not function the same as its conventional counterpart. Remember that the smart transmitter is factory categorized, which means that the reference input pressure is compared to the known output signal.

This information is stored in the transmitter's EPROM (erasable, programmable read-only memory) and is looked at and compared to when the transmitter is in operation. This simply means that at some time, perhaps in the factory, the reference pressure versus output signal has been set. When the transmitter is in operation, the transmitter uses this stored information to produce an output signal in engineering units that corresponds to the input reference pressure. Whether linear or nonlinear, the digital trim allows the user to make corrections to the factory-stored "curve" that is maintained in the EPROM. Digital trim is a two-step process consisting of sensor trim and adjusting the output electronics. The function of digital trim should not be confused with re-ranging because re-ranging does not affect the way a transmitter "sees and interprets" a process input.

Sensor trim exacts the digital process variable to a precision pressure input and allows the user to change the way the transmitter interprets the input variable. To understand digital trim, picture a transmitter with 50 PSI input applied (compared to a knowngauge pressure), but the transmitter sees a slight difference in the input variable. The reason is the factory-stored "curve" was set incorrectly so that the referenced input variable would indicate the correct value. Digital trim corrects that difference by one of two methods: full and zero trim.

A full trim is a two-point calibration procedure in which the lower reference value is applied first. The lower-range value input variable is applied and allowed to stabilize. The upper-range value is applied and the signal is linearized between them. Remember—a calibration procedure is only as accurate as the equipment used. Rosemount recommends using a input reference at least three times as great as the 1151 DP.

A zero trim is a much simpler, one-point process comparison. Perhaps a second pressure is not available, so the "zero" is set accurately instead of a full trim. Simply apply a reference variable that corresponds to a process variable of zero value and set. The reference applied must be within 3% of the true zero stored in the EPROM. If an error is displayed, such as "measurement variable too high," the trim procedure alone cannot calibrate the device. Characterization is necessary, but you should never perform this procedure without first obtaining authorization. This procedure is best performed when the transmitter is installed in its final working position. This allows the user to perhaps gain an advantage by field-setting a transmitter to a known zero reference. When performing a zero trim, the curve used for the range specified is not changed but merely "offset" to correspond to the working zero wanted.

13.6 Output Calibration

Output calibration is another form of the trim procedure. Adjusting the output electronics is sometimes required when a transmitter has been installed and in use for a period of time. To determine if output correction is needed, connect a milliamp meter of known sufficient accuracy and the HART communicator. Perform a loop test (remember to put loop in manual) and set the output to 4 mA. The transmitter output should be within plus or minus 3 microamps.

Next set the device output to 20 mA. The value displayed by the communicator should also be within plus or minus 3 microamps of the measuring milliamp meter. If the tolerances are too great, you must trim the output. Use a current meter as your output standard to read the output units in milliamps and adjust the output to meet the standard for the site where you are working. Trimming the analog output sets the output signal to equal what the reference measuring equipment indicates the transmitter should indicate.

For example, a transmitter may display through the communicator that its output is 20.0 mA, but the output standard indicates a reading of 19.75 mA. Begin the output trim procedure and follow through the steps displayed on the communicator. The output of the device displayed through the communicator and the output standard readings should match when the procedure is complete.

Again, calibrating a smart transmitter is no different than calibrating a "dumb" one. A five-point check is still implemented, the input to the transmitter is still compared to an input standard, and the output signal of the transmitter is still compared to an output standard. When all the values are within tolerances as specified during the five-point check, the device can be said to be calibrated. Do not be misled—calibrating a smart device is not simply entering the ranged values that the device is supposed to measure. You have to verify through the use of measuring equipment (input and output standards) to prove that all values read into and signaled from the device are accurate.

There are also zero and span buttons located inside the cover of the smart electronics housing (Figure 13–5). These buttons allow a user to set the zero and span ranges and settings by using a reference pressure and performing the same calibration procedure as with a conventional transmitter. The smart's zero and span buttons are depressed to perform the same functions as the zero and span screws on a conventional transmitter. When a known, accurate reference variable is connected, depress the span and zero buttons simultaneously for at least 5 seconds to activate the controls. The controls are active for 15 minutes and must be activated again after 15 minutes. Apply a zero reference variable and press the zero button for 5 seconds. Verify the output is at 4 mA and do the same for the high reference point by using the span button.

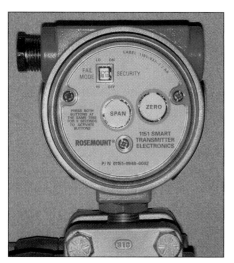

FIGURE 13–5 Zero and span push buttons for the Rosemount Model 1151.

13.7 Multi-Drop Configuration

Sometimes transmitters are connected to transmit an output signal in the digital form rather than the analog form we are used to (Figure 13–6). This does not mean that a transmitter connected to transmit digitally cannot represent an analog range. A transmitter that communicates dig-

FIGURE 13–6 Multi-drop configuration.

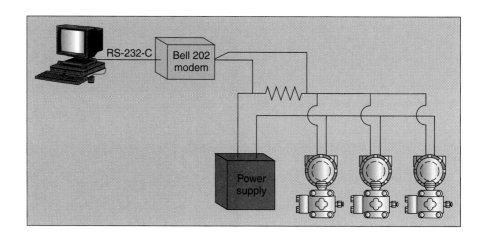

itally is, through a series of discrete bits, able to translate and interpret information.

Multi-dropping transmitters refers to the connection of several transmitters to a single communications transmission line. Communication between the host and the transmitter will occur digitally and the output is deactivated (with Rosemount set to 4 mA). There are limits to the number of devices that are installed on a single communications media and the manufacturer's specifications should be referenced.

When multi-dropping transmitters, each device connected to the communications media must have an address. It is essential that a user verify which functions are deactivated (alarming, 4–20 mA, etc.) when

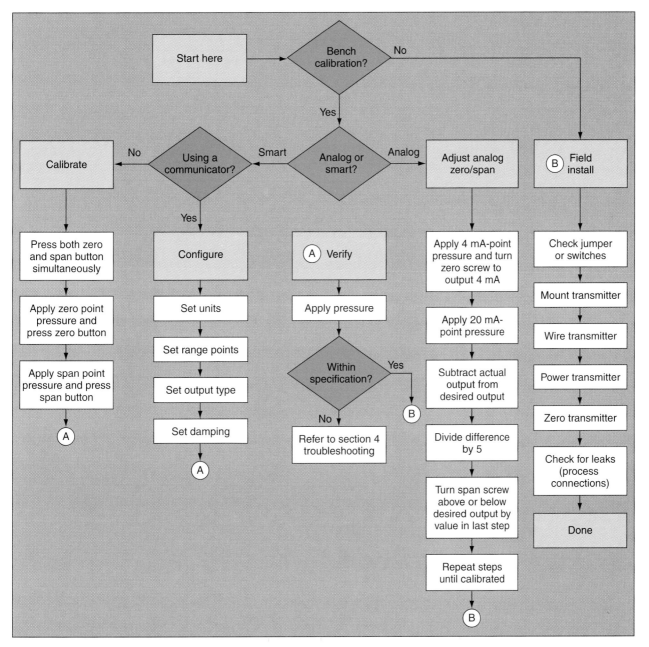

FIGURE 13–7   Calibration procedure(s) flow chart.

re-addressing a transmitter. To set a multi-dropped transmitter, the input reference variable must be known. The output is set digitally to represent the known variable and the host excepts the digital form as the reference variable.

Remember to download configuration data when it is required to give a transmitter working parameters if working off-line. If working on-line, the transmitter will need to be updated.

■ SUMMARY

The smart calibration procedure is required for all smart devices used in process industries. This chapter provided the variables to be concerned with when calibrating a smart device. This chapter also presented the requirements and error checks that should always be used to eliminate instrument calibration errors. This chapter also presented the required procedures that are used to eliminate the errors found in a smart device calibration.

REVIEW QUESTIONS

1. If a device is selected manually to fail low through the use of select switches contained in the smart device, when will the device send a signal lower than 4 mA?

2. In order for communication and configuration to be enabled, should the security switch selection be on or off?

3. Why it is necessary to utilize test equipment to verify a calibration of a smart device as compared to using a smart communicator?

4. During a calibration procedure, a smart device indicates that it is reading an input pressure of 97.5 $''H_2O$, but the certified calibration input standard indicates an input pressure of 100 $''H_2O$ is supplied to the transmitter. Which value should be taken as accurate, and what procedure would be used to correct the error? What should be the final result?

5. During a calibration procedure, a smart device indicates that it is signaling an output signal of 17.4 mA, but the certified calibration output standard indicates an output signal of 18.0 mA. Which value should be taken as accurate, and what procedure would be used to correct the error? What should be the final result?

chapter 14

Fundamentals of Instrument Installation

■ OUTLINE

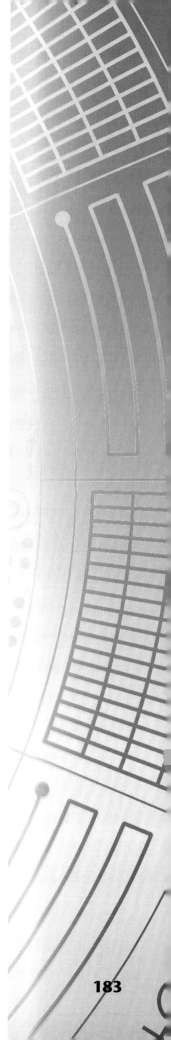

■ OVERVIEW

Instrument installations are a possible source of error. An incorrect installation can present maintenance, calibration, and physical hazards. This chapter provides the necessary criteria to explain the proper requirements for device installations. After you read this chapter, you will understand the correct procedures for mounting, tubing installation, wiring, and grounding.

■ OBJECTIVES

After completing this chapter, you should be able to:

- List the required areas of concern for installing an instrument.
- Describe the proper installation methods for process tubing and connections.
- Install the required wiring and signal path requirements.
- Correctly route, install, and seal conduit used for instrument signals.

■ INTRODUCTION

14.1 Fundamentals of Instrument Installation

The differential pressure type is one of the most widely used field transmitters (Figure 14–1). We know from our studies that the differential pressure transmitter performs by sensing the differential pressure of a process to obtain flow, level, pressure, and, in some cases, temperature to sense a process change. Our discussion here will show the steps that are used to install a differential pressure-type transmitter. These steps can be applied generally to all devices used in a "normal" process environment.

14.2 Installation—Mounting

Instrument installation consists of three separate functions: mounting, tubing, and wiring. Each function should be performed so as to cause a minimum amount of process deviation as well as signal value.

Sometimes installers have little choice over the mounting position, wiring, and tubing while, at other times, they may have a choice for all three. Each separate part of an instrument's installation is important to the overall function of the device.

The accuracy of any field-mounted transmitter depends on the installation of that device (Figure 14–2). If a transmitter is mounted in a position that changes the actual process pressure that it is measuring, the overall goal of maintaining an efficient, accurate, and automatic process cannot be achieved. Likewise, if the wiring is run in a way that introduces stray signals, the end result is the same—an incorrect process variable.

Once the proper position for mounting with respect to the impulse tubing is chosen, the installer will also need to check other considerations. Accessability, safety for field personnel, ease of field calibration, and a practical working environment are other considerations a field installer should consider.

Installations in certain process environments may cause the need for special seals and other installation requirements to satisfy regulations, such as those in the food, beverage, and pharmaceutical industries. In short, the requirements for specific site installations will be dependent upon the location where the instrument is to be installed.

FIGURE 14-1 Dimensional drawing for Rosemount Model 1151.

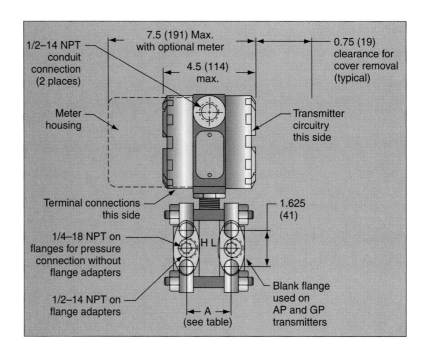

FIGURE 14-2 Dimensional drawing for Rosemount Model 1151.

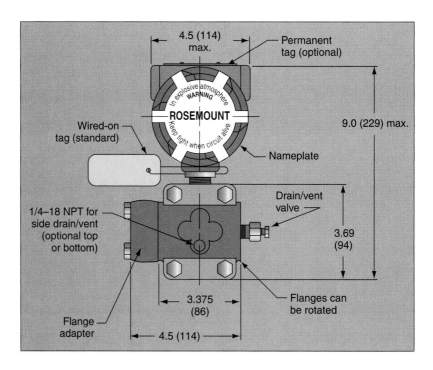

Specific requirements for instrument installations should be available from the worksite. However, there are general requirements for all instrument installations.

In general, transmitters should be installed to minimize vibration, shock, and temperature fluctuations. Most transmitters can be mounted in one of three ways: wall-mounted, panel-mounted, or attached to an instrument (pipe) stand (Figure 14–3). Regardless of the mounting method, you must take into account the ease of getting to the

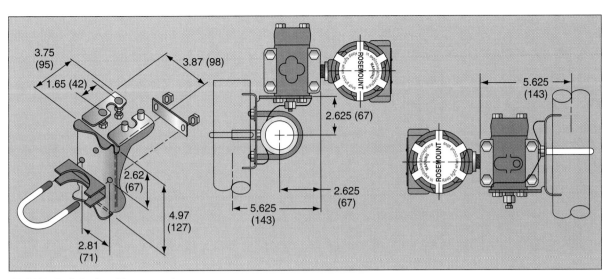

FIGURE 14-3 Pipe stand mount.

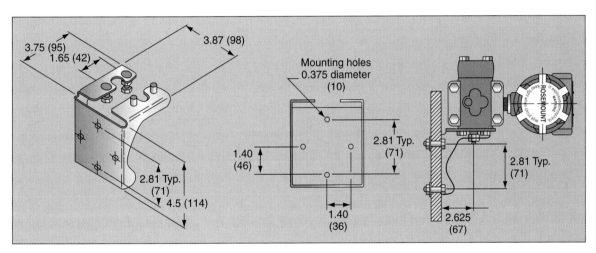

FIGURE 14-4 Panel mount.

transmitter's electronics as well as its field-wiring terminations, and do not forget the zero and span adjustments. You want to consider the process connections and the drain/vent valves on the transmitter so that when they are opened, they will be directed away from the field technician.

The electronics housing can be rotated no more than 90° when mounting. If you rotate the housing more than 90° you may damage the internal sensor wiring. To rotate the housing less than 90°, you must loosen the housing lock nut and turn the housing. Don't forget to tighten it when you finish. If the housing needs to be turned more than 90°, the transmitter must be disassembled and reassembled with the housing correctly positioned (Figure 14–4).

Once the process side has been established, care must be taken so that proper access is available for the electronics side. Only 3/4 inch is required to remove most electronics housings, but 6 inches is preferable. The electronics side is not generally opened once installed, but care

FIGURE 14–5 Flat mount.

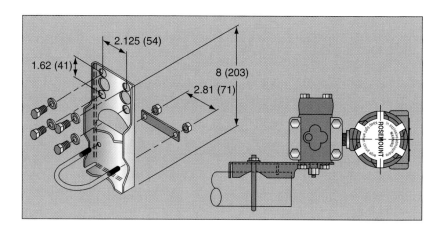

should be given to ensure its ease anyway (Figure 14–5). Smart devices often require access to both sides of the device. Wiring connections are to be made through the conduit openings located on the top of the housing.

14.3 Installation—Tubing

We learned earlier that the mounting position of the transmitter must be the same as the calibrated position. For electronic devices, the heat dissipation of the electronics can cause slight variations in the transmitter's reading. When a differential pressure transmitter that is measuring liquid level is mounted in a position other than that calibrated, the sensor will record a slight difference also. This difference is equivalent to the amount of "zero" shift, which is the amount of the "head" change that is introduced from shifting the sensor position with respect to gravity.

Different mounting requirements depend on the type of service the transmitter is to record. Steam, liquid, and gas are the three types of process we may have to measure. For liquid measurement, you should mount the transmitter such that the process-measuring taps are to the side of the line so that sediment does not become trapped. Mount the transmitter to the side or below the taps to ensure that air does not become trapped in the impulse tubing (Figure 14–6)

For measuring gas flows or gas levels, the process taps should be located in the top or the side of the line to ensure that liquid will drain into the process line. For steam measurement, the process taps should be located in the side of the line and the transmitter should be mounted below the taps to ensure the process tubing remains filled with condensate (Figure 14–7).

Most transmitters are equipped with drains and/or vent valves, so the process measured determines how these valves are aligned. If measuring liquid, the valves should be mounted upward of the line to allow gases to vent. If measuring gases, the valves should be mounted to allow any liquid that has collected in the process tubing to drain. Also, if measuring steam, the valves should be mounted the same as those for liquid measurement because we know now that the lines should be filled with water. Filling the steam service lines with water is required to prevent steam from coming into contact with the transmitter sensor (Figure 14–8).

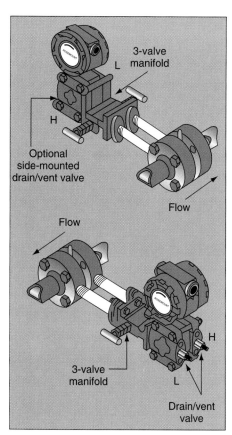

FIGURE 14–6 Liquid service.

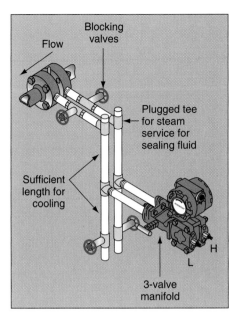

FIGURE 14-7 Steam service.

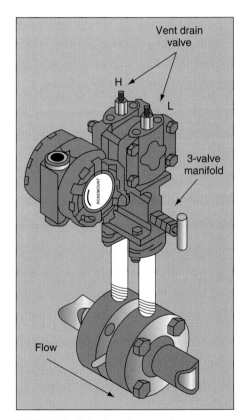

FIGURE 14-8 Gas service.

Remember the relationships studied earlier—the potential zero shift due to excessive pressure is provided to the sensor of the transmitter. Elevation and/or suppression are two conditions that an installer must consider when measuring liquids. For special cases of gas measurement, elevation and suppression may be considerations but we now must consider a process that is lighter than the relationship we are used to dealing with. Mounting details should provide all necessary requirements for mounting height with respect to tap positions.

It is possible that gas-measuring devices can be mounted to show an excess or lack of process due to the relationship of the process to the mounting position. Mounting details should be provided for special-service devices to ensure that they are not located improperly.

The piping (tubing) between the process and the transmitter should remain at the shortest length possible. Using a minimum of impulse piping ensures that the accuracy of the device is not decreased. When installing a device, the installer must consider the field-impulse tubing route as well as the electrical signal wiring route.

Impulse tubing carries the signal from the process to the transmitter. The tubing must be of the shortest length possible. The installer must also consider the elevation changes between the process and the transmitter. If the tubing elevation changes, the process pressure sensed by the transmitter can be of a different value than the actual process that should be measured. Also, if the tubing is routed in a way that allows moisture to accumulate inside the tubing, the process pressure can be changed. The following rules, when followed, should alleviate most problems associated with the transmitter tubing and mounting location:

1. Process tubing length should be as short as possible.
2. Process tubing should be sloped upward from the transmitter to the process for liquid measurement and downward from the transmitter to the process for gas measurement.
3. Tubing should be routed to avoid "high spots or crowns" in a liquid measuring line.
4. Tubing should be routed to avoid valleys where liquids can collect when measuring gases. Sometimes drains are required for low points to drain any liquids that may accumulate (Figure 14–9).

Both process tubing legs (if used for flow measurement) and closed-vessel level measurement should remain the same temperature and, preferably, the same length. For low points in gas tubing, a drain for gases must be present for liquid lines, just as with liquid drains. The temperature change can often introduce condensation into a line that measures gases. If "wet legs" are used, the same level of liquid must be supplied to both legs. Keep sediment deposits out of liquid-filled process tubing lines.

Sometimes it is necessary to purge a tap connection to a process to prevent the tap from plugging. Be sure to make the purge connection as close to the taps as possible. Try to avoid purging through the transmitter. On a related note, all hot and/or corrosive process materials must be kept from coming into contact with the sensor of the transmitter.

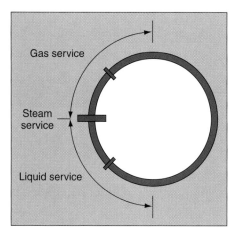

FIGURE 14–9 Process line cross-section indicating tap position(s).

FIGURE 14–10 LCD display.

FIGURE 14–11 LCD installation; exploded view.

We now know the general rules for running process tubing and for determining the correct mounting location for the instrument. Wiring considerations must also be made to allow for efficient and accurate process measurement signals. Remember, it is the electrical signal that transmits the process variable to a controller. All other considerations may have been made correctly and still an error is introduced on the signal side of the transmitter and not on the sensor side. The following wiring example is for the Rosemount Model 1151 DP transmitter, but the guidelines can be used for practically all devices.

14.4 Installation—Wiring

There are two sets of terminals located in the compartment that is opposite the electronics housing. The signal terminals are the upper terminals and the test terminals are the lower set. The signal wiring is connected or terminated at the signal terminals. Simply connect the wiring conductor that originates from the positive side of the loop's power source to the positive terminal and the negative conductor to the negative terminal. The transmitter is often the last device mounted and, when mounted, the loop is completed and a signal should be able to be transmitted. Remember that the hazardous area classifications exist for instrument signal wiring. Observe the same precautions you take for electrical power wiring.

The test terminals have an output for a 4–20-mA loop that is identical to that of the signal terminals. The only wiring that should be connected to any test terminal is a integral meter for the device and you can use the test terminals for reading the current of a 4–20-mA loop (Figure 14–10). If an integral display meter is used, follow the mounting instructions carefully. Individual device types will vary (Figure 14–11).

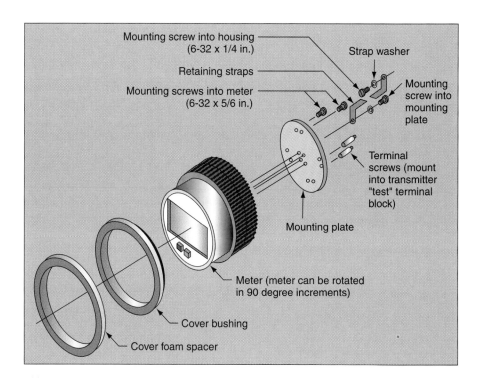

FIGURE 14–12 Field-wiring path, grounding.

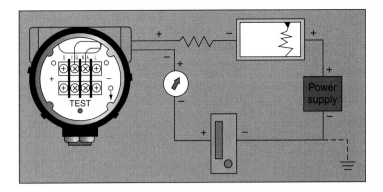

To measure loop current, simply place the leads of a meter positive-to-positive with the test terminals and negative-to-negative. You should be able to measure a current. Remember—do not connect power to the test terminals. Connecting a voltage to the test terminals may burn out a biasing diode, which will not allow any local indication of the current in the loop.

Signal wiring, as we know, carries the 4–20-mA loop current that is used by a device or controller that will indicate the process measured. Care must be taken when routing signal wiring that it does not "pick up" stray voltages or noise. If signal wiring is routed in conduit or open trays near heavy power loads or high voltages, it is easy to see that a difference in the signal strength can be added to the current loop.

Most methods for signal wiring incorporate the use of a "twisted pair with shield." This implies that the signal conductors are twisted together to cancel any magnetic field effects. The shield also cancels any capacitive effects that may be present.

The shield of the signal wiring can be left ungrounded, but some applications may call for a grounded shield. The shield should be grounded at the source end, which means that whatever is supplying power for the transmitter and resulting signal current is the source end. The negative terminal of the source end is a good point for grounding. Wiring diagrams show in detail where the shields should be grounded (Figure 14–12).

In addition, the transmitter case must be grounded. It is most likely that sufficient grounding was provided when the transmitter was mounted, but a direct connection from the case to a grounding electrode can be made if necessary. It is important to note that the loop-wiring sheets should also indicate where a loop is to be grounded, if at all.

14.5 Installation—Conduit

Consider the conduit run (if conduit is used) for the signal wiring. Conduit should be run to allow for maximum clearances around potential noise sources. Do not parallel high-current/voltage lines if possible. Do not terminate the conduit so as to provide a channel for moisture into an electronics enclosure.

Conduit seals are recommended when terminating to a field transmitter (Figure 14–13). Unused conduit openings should be plugged and sealed. 4–20 mA analog signals are easier to route conduit for than, say, a temperature loop that is concerned with reading an analog range

FIGURE 14-13 Conduit installation details.

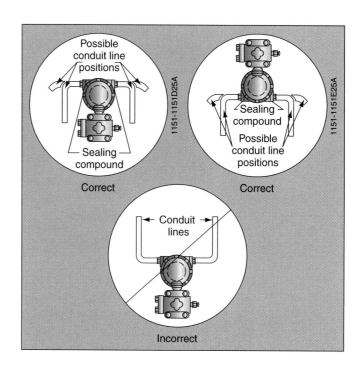

FIGURE 14-14 Model 3051 PT with LCD display.

of millivolts. Also, for a digital signal, such as that used for multi-dropped transmitters, extra care must be taken to observe noise sources as well as the wiring methods used. In some cases, the category 5 or later wiring methods have to be installed as per digital communication standards. Again, this should be specified on the associated documentation sheets.

Lastly, conduit runs that contain signal loops should be run so that high-frequency switching is not impressed upon the signal wiring. If a high-frequency switching load is present in the area, the conventional wiring (4–20 mA) loop should not be run in the same conduit as the high-frequency switching source. It is acceptable, in most cases, to run multiple analog signal loops in the same conduit if shielded twisted pairs are used.

14.6 Practical Examples

Figures 14–14 through 14–18 provide examples of installation requirements and field adjustments to signals. Each figure shows a proper installation given the process and measurement ranges.

Figure 14–19 shows the installation of an opto-isolator. This device physically separates a wiring loop signal path while regenerating the same signal value. For example, a device is field-mounted and sends a signal to the controller, but there is a difference of potential from the field device and the controller due to a difference in ground potential. This may cause an incorrect signal to be received by the controller, so an isolator is installed to remove the error.

FIGURE 14–15 Steam measurement; notice steam pots to keep tubing full, isolation valves, and three-valve manifold.

FIGURE 14–16 Venturi differential measurement. Observation of the tubing route, mounting location, and isolation valves indicates what type of service: liquid, gas, or steam?

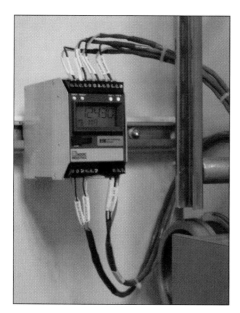

FIGURE 14–17 Converter of digital to analog signals from 3095 MV transmitter to read temperature, pressure, and differential pressure.

FIGURE 14–18 3095 Multi-variable transmitter, inserted pro-bar to calculate DP, pressure, and temperature.

FIGURE 14–19 Opto-isolater.

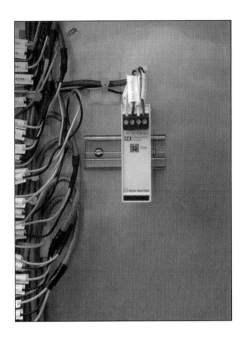

FIGURE 14–20 Oxidizer field panel with seal-offs. Notice pressure transmitter at the bottom to monitor panel positive pressure.

■ SUMMARY

Installation requirements often are dictated by the details portrayed on drawings that are provided. This chapter did not try to circumvent that process, but rather provided the concerns that should be present for all installations. This chapter provided installation requirements related to wiring, tubing, and conduit and discussed the reasons why such concerns should be addressed.

■ REVIEW QUESTIONS

1. Give the three separate functions that are required to install an instrument correctly.

2. In general, devices are installed to minimize three areas of concern. What are they?

3. For liquid differential flow measurement, should the transmitter be mounted above or below the process tap connections? Explain your answer.

4. Why should a steam-measuring device have its process tubing legs filled with condensate (water)?

5. If voltage is applied to the test terminals of a device, what might be the result?

6. Grounding (shield) is usually terminated at one end. Which one, and why?

chapter 15

Fundamentals of Instrument Maintenance

■ OUTLINE

▣ OVERVIEW

This chapter provides information on the types of errors that occur in process control devices, and their resulting corrections. The chapter does not address each error, but rather provides valuable assistance in troubleshooting skills so all errors can be identified. The chapter discusses problem solving, troubleshooting, and diagnostic tools for both smart and conventional devices. Device electronics and sensor checks and repairs are also explained.

▣ OBJECTIVES

After completing this chapter, you should be able to:

- Provide field assistance when troubleshooting devices.
- Discuss and perform troubleshooting diagnostics on a smart device.
- Discuss and perform troubleshooting procedures on a conventional device.
- Determine the cause of error signals and correct them.
- Provide knowledge of device and control-loop analysis.

▣ INTRODUCTION

15.1 Fundamentals of Instrument Maintenance

Instrument maintenance in a working environment is often dictated by a set procedure. A preventative maintenance (PM) schedule is often adopted by a "working" site as a possible security measure to detect faulty devices.

The process that is under control, along with the location of devices, will determine the extent of any "on-line" maintenance procedures. On-line testing procedures test a device for errors while it is recording and transmitting process variables.

The type of device under inspection also determines the maintenance procedure that is followed. Smart devices require different procedures than those for conventional devices. There is also a thin line between maintenance and troubleshooting. Often enough, PM schedules will incorporate predetermined troubleshooting steps into the procedure for particular devices.

Set procedures that establish certain PM steps to repair/replace/rebuild are much more common for conventional devices for several reasons. Conventional devices have been around longer and, therefore, maintenance personnel are more familiar with them. Also, there are more conventional devices than smart devices installed (although this is changing rapidly) and, therefore, they are more likely to have an established troubleshooting procedure. Lastly, and more importantly, a smart device can often do a simple, self-check procedure to find possible trouble spots.

Regular PM schedules for smart and dumb devices often are written to test a device for accuracy, functionality, signal proportions, repeatability, etc. If a device does not meet the criteria that are given in the PM procedure, it is often up to the person performing the procedure to determine the cause.

The troubleshooting guidelines given in this chapter are no substitute for field experience. Field technicians who are performing troubleshooting procedures must be able to recognize a faulty device. Often, a device records and transmits a correct signal for the process, but it still is not performing correctly. It may take too much time to reach its

> ### ⚠ WARNING
>
> Process leaks can cause death or serious injury. Only use bolts supplied with the transmitter or sold by Rosemount Inc. as a spare part. Using unauthorized bolts may reduce pressure retaining capabilities and render the instrument dangerous.

> ### ⚠ CAUTION
>
> Do not plug the low side with a solid plug. Plugging the low side will cause an output shift.

> ### ⚠ WARNING
>
> Explosions can cause death or serious injury. Do not remove the instrument cover in explosive atmospheres when the circuit is alive.

> ### ⚠ CAUTION
>
> Do not connect the power signal wiring to the test terminals. Voltage may burn out the reverse-polarity protection diode in the test connection. If the test diode is destroyed, then the transmitter can still be operated without local indication by jumping the test terminals.

output signal compared to process change; it may transmit "spikes" (harmonic or electronic distortions) in its output signal; it may sense a process change through a partially plugged sensor or require a longer time to sense the process change because of a faulty sensor seal. These examples are only a few of the many inaccurate signals that can be transmitted by a faulty transmitter.

Often enough, there is nothing wrong with a device that is suspected of causing a problematic signal. The problem may lie in related field wiring, termination points, configuration in controllers, location, etc. Faulty signals can also be the result of a weather change that affects the working environment and reference pressures of instruments.

The following sections separate the troubleshooting procedures for smart and conventional devices, but you will find several similarities. The guidelines present possible steps that you can take to discover a faulty process variable reading. The first step (if possible) is to remove power source(s) and supply pressures. Also, block any process fluids from the sensor until they are needed.

15.2 Smart Device Troubleshooting

Faulty Communications

To perform diagnostic procedures to a smart device, a communicator must be connected to the loop wiring. Sometimes the communicator cannot "see" the device and troubleshooting steps must be taken to establish communications. Potential causes of a communications failure generally exist with the loop wiring, in which there are several possible explanations for why a device is not communicating: There may not be the 250 ohms of minimal loop resistance present; or there may be less than the minimum of 17 volts present for the communicator to operate.

To address these problems, measure the loop voltage to see if possible shorts and/or open circuits are preventing a continuous operating voltage. Multiple ground points could be inducing loop currents that affect the signal between the communicator and the smart device. Also verify that the device you are trying to communicate with is a smart device. You will not be the first troubleshooter who has written a device inspection procedure calling for a smart device maintenance on a dumb device. You can eliminate a lot of related field-wiring problems by connecting directly at the transmitter and then at a remote location and comparing the response of the communicator to the two locations.

Another related troubleshooting point for loop wiring occurs at any intrinsically safe wiring barriers that are in the loop wiring. Barrier troubleshooting documentation is the only way to determine if a barrier is performing correctly. You must follow the procedure(s) correctly to determine if any error measurements are in the barrier.

Consistent High Output

Consistent high output is shown in Figure 15–1. Once communications are established, you can then isolate and diagnose the trouble

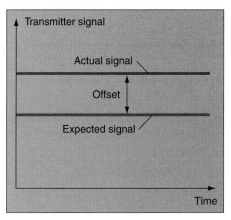

FIGURE 15-1 Transmitter signal response, consistent elevated signal.

> ⚠ **WARNING**
>
> The following performance limitations may inhibit efficient or safe operation. Critical applications should have appropriate diagnostic and backup systems in place. Pressure transmitters contain an internal fill fluid. It is used to transmit the process pressure through the isolating diaphragms to the pressure sensing element. In rare cases, oil leak paths in oil-filled pressure transmitters can be created. Possible causes include physical damage to the isolator diaphragms, process fluid freezing, isolator corrosion due to incompatible process fluid, etc. A transmitter with an oil fill fluid leak can continue to perform normally for a period of time. Sustained oil loss will eventually cause one or more of the operating parameters to exceed published specifications while a small drift in operating point output continues. Symptoms of advanced oil loss and other unrelated problems include:
> - Sustained drift rate in true zero and span or operating point output or both
> - Sluggish in response to increasing or decreasing pressure or both
> - Limited output rate or very nonlinear output or both
> - Change in output process noise
> - Noticeable drift in operating point output
> - Abrupt increase in drift rate of true zero or span or both
> - Unstable output
> - Output saturated high or low

spots. Suppose the signal of a device is consistently higher than it is supposed to be. Of course, you should verify the calibration first and, if errors still exist, then check for restrictions at the primary element such as crimped tubing, closed valves, etc. Check the process tubing for leaks or blockages. Verify that all blocking/isolating valves are fully open. Bleed all of the process tubing (if allowed) to ensure that gases are not present in the tubing. Ensure that the density/specific gravity of the fluid has not changed. Clean out any sediment that may have accumulated in the process flanges/taps.

You should check the power supply output to verify the voltage at the device is in the operating range. The internal workings of the smart transmitter electronics depend upon an established range for a working voltage. At any voltage outside that range, the transmitter may continue to operate but with faulty readings and/or signals.

The transmitter electronics can also be a source of errors even if the working voltage is within the working range. The transmitter test mode should be accessed and executed to find any errors with the electronics. Sometimes intermittent results are obtained from a transmitter test command given from the communicator. This could also be the same cause as the one that is giving a high output. Clean the electronic post connectors to ensure a constant communication path is available and verify the transmitter response again.

If a problem still exists after you have checked the process tubing, measured the power supply(s), and checked the transmitter electronics through the use of the transmitter test procedure of the smart calibrator, the trouble may lie within the sensing element of the device. Sometimes a visible leak in a sensor is detectable and sometimes it is not. Regardless, the sensor is not repairable while it is in service. The device must be removed to disassemble and repair if possible (to be covered later).

Consistent Low Output

Consistent low output is shown in Figure 15-2. The same points of interest for a high output should be inspected for a low output. Verify calibration first and then begin troubleshooting. Check the primary element for restrictions.

Verify that all process tubing is in good working order. Check the process's specific gravity to ensure proper calibration ranges. Sediment and/or gases trapped in the lines could alter signal proportions. Blocking valves could prohibit the process from being accurately measured if one is partially closed.

Verify that the loop wiring is correct by checking the working voltage range. Eliminate multiple ground points. Check the loop impedance to see if it is too high and limits the output signal. An often overlooked trouble spot is power source overloading. An overloaded power source cannot supply adequate current to all of the devices connected to it. Compare a power supply's total milliamp output to the total of all devices connected to the source. A common mistake is to assume that if a power supply is overloaded, all devices will be affected—this is not always the case. A power supply will normally be

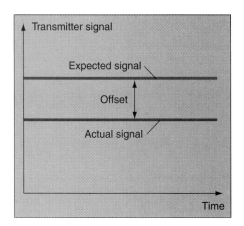

FIGURE 15–2 Transmitter signal response, consistent suppressed signal.

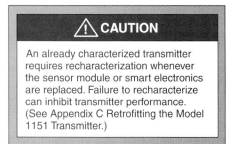

⚠ **CAUTION**

An already characterized transmitter requires recharacterization whenever the sensor module or smart electronics are replaced. Failure to recharacterize can inhibit transmitter performance. (See Appendix C Retrofitting the Model 1151 Transmitter.)

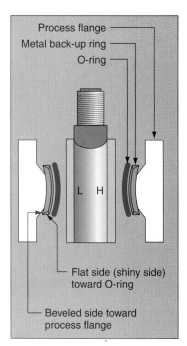

FIGURE 15–3 Detail showing process O-ring and back-up ring installation.

current-limited, which means that the power supply will not be able to output enough current to satisfy the requirements of all of the supplied devices. By recalling Ohm's law and Kirchoff's laws, we know that only the loop with the most resistance will be affected first and then others will follow as the load is increased. This not always the case but it is good to remember as a possibility.

In a device that is outputting a low signal (or no signal), you should check for the polarity at all termination points. Perform an insulation test of the signal conductors to verify a short is not causing a bleed-off of some of the signal's strength.

A transmitter that outputs a constant 4 mA when hooked up and put on-line may be in the multi-drop mode. Recall that the multi-drop mode of a Rosemount transmitter is field selectable. A previous installation/calibration procedure may have resulted in an inadvertent switch setting in the multi-drop mode. A communicator loop test can find a lot of the loop wiring problems, but remember to put all loops that may be affected into manual mode.

You should test the transmitter electronics. Initiate a transmitter test from the communicator to find any electronics failures. If you find no error, and the electronics are still suspect, you can replace the electronics (to be covered later). Finally, the cause of the error may be in the sensing element (Figure 15–3). The element is not generally field repairable and must be replaced (to be covered later).

Other Error Modes

Other error(s) may be present in a smart device's signal, such as an erratic output or a transmitter that does not record the process except in segments of the working range. By checking and verifying the methods of troubleshooting given previously, a field technician can eliminate most of them.

The advantage of the smart device is the hand-held communicator, which can save you a lot of time when you troubleshoot devices. The functions of loop test and transmitter test should be used to their fullest potential. Some industrial sites require the transmitter test and loop test commands to be performed on a regular schedule to verify a working process loop integrity. You should always perform such actions with the acknowledgment of the proper personnel. Take care to observe the working order of manual and automatic process loops that can possibly be influenced.

Sometimes an intermittent error is observed for which there is no quickly discernable answer. Field hands often refer to such errors as "unidentifiable." That may be the case for the first time. If you are present when such an error is observed, record the environmental conditions, process working conditions, other work (maintenance/construction) that may be occurring, alarms that may have been received at the same time, and other devices and/or loops that may have been affected, etc. Chances are, as time goes on, the "unrelated" conditions may be shown to be related after all. This is the true meaning of having good field experience.

15.3 Conventional Device Troubleshooting

Consistent High Output

Verify the calibration first and, if errors still exist, check for restrictions at the primary element such as crimped tubing, closed valves, etc. Check the process tubing for leaks or blockages. Verify that all blocking/isolating valves are fully open. Bleed all of the process tubing (if allowed) to ensure that gases are not present in the tubing. Ensure that the density/specific gravity of the fluid has not changed. Clean out any sediment that may have accumulated in the process flanges/taps.

Check the power supply output to verify the voltage at the device is in the operating range. The internal electronics of the transmitter depend on an established range for a working voltage. At any voltage outside that range, the transmitter may continue to operate but with faulty readings and/or signals.

The transmitter electronics can also be a source of errors even if the working voltage is within the working range. Clean the electronic post connectors to ensure a constant path is available and verify the transmitter response again.

If a problem still exists after you have checked the process tubing, measured the power supply(s), and checked the transmitter electronics (connected), the trouble may lie within the sensing element of the device. Sometimes a visible leak in a sensor is detectable and sometimes it is not. Regardless, the sensor is not repairable while it is in service. The device must be removed to disassemble and repair if possible (to be covered later).

Consistent Low Output

The same points of interest for a high output should be inspected for a low output. Verify calibration first and then begin troubleshooting. Check the primary element for restrictions.

Verify that all process tubing is in good working order. Check the process's specific gravity to ensure proper calibration ranges. Sediment and/or gases trapped in the lines could alter signal proportions. Blocking valves could prohibit the process from being accurately measured if one is partially closed.

Verify that the loop wiring is correct by checking the working voltage range. Eliminate multiple ground points. Check the loop impedance to see if it is too high and limits the output signal. An often overlooked trouble spot is power source overloading. An overloaded power source cannot supply adequate current to all of the devices connected to it. Compare a power supply's total milliamp output to the total of all devices connected to the source. A common mistake is to assume that if a power supply is overloaded, all devices will be affected—this is not always the case. A power supply will normally be current-limited, which means that the power supply will not be able to output enough current to satisfy the requirements of all of the supplied devices. By recalling Ohm's law and Kirchoff's laws, we know that only the loop with the most resistance will be affected first and then others will follow as the load is

FIGURE 15–4 Exploded view of model 1151E.

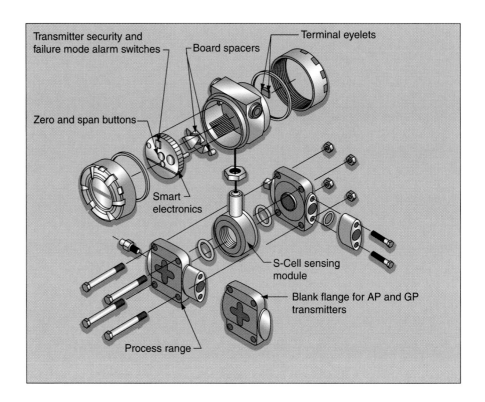

increased. This not always the case but it is good to remember as a possibility.

In a device that is outputting a low signal (or no signal), you should check for the polarity at all termination points. Perform an insulation test of the signal conductors to verify a short is not causing a bleed-off of some of the signal's strength.

Test the transmitter electronics to see if they are connected properly. If you find no error and the electronics are still suspect, you can replace the electronics (to be covered later). Finally, the cause of the error may be in the sensing element. The element is not generally field repairable and must be replaced (to be covered later; see Figure 15–4).

Other Error Modes

Other error(s) may be present in a conventional device's signal, such as an erratic output or a transmitter that does not record the process except in segments of the working range. By checking and verifying the methods of troubleshooting given previously, a field technician can eliminate most of them. Elevation and suppression errors may need additional adjustments and require the device's component board to be adjusted (Figure 15–5).

The advantage of the conventional device is that it is more familiar to most field personnel. Most likely, one of your coworkers has already experienced the same problem(s) that you may be having.

As with smart devices, sometimes an intermittent error is observed for which there is no quickly discernable answer. Field hands often refer to such errors as "unidentifiable." That may be the case for the first

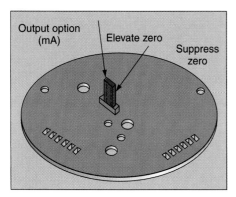

FIGURE 15–5 Elevation and suppression using electronics.

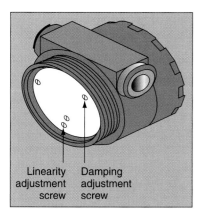

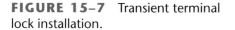

FIGURE 15–6 Linearity and damping adjustments.

time. If you are present when such an error is observed, record the environmental conditions, process working conditions, other work (maintenance/construction) that may be occurring, alarms that may have been received at the same time and other devices and/or loops that may have been effected, etc. Chances are, as time goes on, the "unrelated" conditions may be shown to be related after all (Figure 15–6). This is the true meaning of having good field experience. More data should be present for the conventional device than for the smart one. Take care not to overlook the database for instrument errors.

Most process environments have a device history list, which is exactly what it sounds like. A device is tracked from its installation through the present and any significant data that reflect troubleshooting, replacement of parts, repair of parts, etc. should be located there. There are very few errors that are actually recorded for the first time. Usually, a segment or, perhaps, the entire error is indicated by the response, or lack of response, of an instrument.

Sometimes a device's errors can be interpreted from the action of other devices that are recording the same process variable. This is often the case with critical control devices that are duplicated and sometimes triplicated. Such multiple sensing and/or final control elements are "voted" and used in controller diagnostics to determine instrument errors and control responses.

The nature of such devices often confuses or makes a field hand extra cautious. For example, a triplicate process variable requires a process variable that is the "median" of the three—that is, the middle value that is recorded from the triplicate set is the one used. If one of the triplicate devices is erratic, the error device will have no bearing on the controller because it will not be the middle-valued device. The thing to remember is that with voted transmitters, safety for process and personnel is the intent behind the multiple sensors and/or final control elements.

Sometimes a device will require transient protection; when this occurs, you should install a transient protection terminal lock (Figure 15–7).

FIGURE 15–7 Transient terminal lock installation.

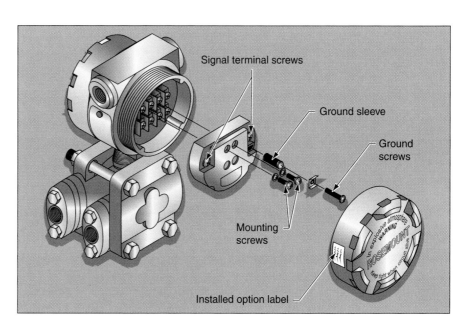

15.4 Sensor Body Removal/Replacement

Our discussions here will be limited to the differential pressure transmitter Rosemount Model 1151. The sensor body is the assemblage that measures the actual process differential pressure.

Remove the transmitter from service and then disconnect the process pressure and power supplies. When you remove the transmitter from service, you must disconnect the process flanges by taking off the four process flange bolts.

Unscrew the cover on the terminal side of the transmitter after verifying that the power is off. Remove the screws and unplug the electronics. Loosen the locknut and remove the standoffs.

You are now ready to remove the sensor module. Unscrew the sensing module from the electronics housing, taking care that the sensor leads do not become damaged. Pull the header assembly board through the hole. The wiring connection is sealed with a sealing compound—you need to break the connection to proceed.

The sensing module is not repairable because it is a welded assembly. Examine the sensor for possible points of damage. There may be a hole in the sensor or there may be burn marks on the sensor from being exposed to a process temperature outside the working range of the transmitter. A loss of fill fluid may be discernable.

If you cannot find an obvious defect, you should check out the sensor module by first removing the header board connection assembly from its post connectors. The sensor module and the electronics can remain attached for the checkout. Check the resistance between the sensor module housing and pins one, two, three, and four. This function checks the resistance between the capacitor plates and ground. The reading should be in excess of 10 M ohms. Measure the resistance between pin eight and the sensor module to ensure that the module is grounded. This resistance should be zero ohms.

These measurements check only for possible shorting or isolating conditions of the sensor. There are other possible errors that may come into play. If a problem persists and you suspect that the cause may be a faulty sensor, then replace the sensor.

⚠ **WARNING**

Failure to install flange adapter O-rings can cause process leaks, which can result in death or serious injury.

There are two styles of Rosemount flange adapters, each requiring a unique O-ring, as shown below. Each flange adapter is distinguished by its unique groove.

Flange adapter O-ring

Unique O-ring grooves

Flange adapter O-ring

FIGURE 15–8 Model 3051 installation detail for differential and gauge pressure measurement.

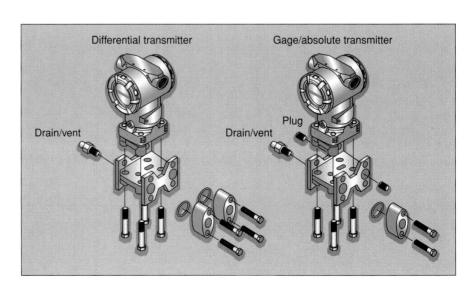

When you reinstall the components, visually inspect the device to verify there are no damaged pieces. O-rings, diaphragms, screws, brass seats, etc. should all be checked out while the device is disassembled and easier to view.

Lightly coat the O-rings with an inert oil. Then insert the header assembly board through the electronics housing. Screw the sensor module into the electrical housing until the threads are fully engaged, being careful not to damage the sensor leads. Align the sensor so it faces the way that is convenient for installation and tighten the locknut.

If a transmitter is of high-pressure type or gauge-pressure type, there is a slightly different procedure for re-bolting the process flanges to the sensor (Figure 15–8). For a standard DP sensor, the bolts are tightened diagonally to the specified torque established by the location where you work. The transmitter is now ready for calibration with a new sensor installed (Figure 15–9).

FIGURE 15–9 Model 3051 exploded view.

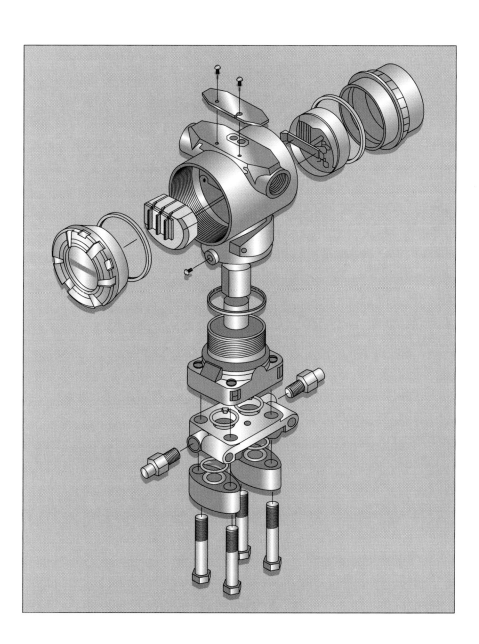

FIGURE 15–10 Hand-held pressure gauge. Test meter applications include using hand-held meters to verify or check device's accuracy.

Characterization is the one-time calibration of the sensor. Recall that a sensor and its accompanying electronics have a series of "curves" that are used to interpret the transmitter's output signal when exposed to a process variable. A transmitter that cannot be "zeroed" must go through a characterization procedure to enable it to read accurately. A transmitter has no curves and must be told how to respond when a pressure differential is sensed.

Characterization of a retrofitted transmitter or a transmitter that cannot be zeroed requires the use of a HART communicator, a pressure source, a milliammeter, and a power supply. It is good practice to "pressure up" the sensing plates on a retrofitted transmitter several times before beginning characterization procedures to ensure the devices installed have been seated properly.

Characterization involves comparing the pressure input to the output of each transmitter's sensor module. Remember that the calibration of a transmitter is only as good as the calibration equipment that is being used. The same standards of accuracy that apply to standard calibration procedures apply here as well.

During the characterization process, you will be required to provide input pressure to the device (for a pressure transmitter; use ohms for a RTD measurement, millivolts for a thermocouple, etc.) and you should use an input standard. The characterization procedure is available through the HART communicator. Follow the procedure exactly as directed, and the result is a "new" transmitter curve built into the device that should allow accurate response. Never perform this procedure without authorization—some manufacturer's warranties are voided if this procedure is performed. The facility you are working in will determine if characterization should be performed.

■ SUMMARY

This chapter provided a list of procedures, devices, and test equipment that are often used for troubleshooting devices that have been operating under process conditions. This chapter did not provide calibration or documentation requirements, for they were covered in other chapters of this text. This chapter did provide procedural, analytical steps to determine if a device is in error and outlined the steps needed to identify the error cause if it is related to the device or impulse (process) tubing.

■ REVIEW QUESTIONS

1. What is the minimum resistance required to use a HART communicator for device troubleshooting, diagnostics, and communication?

2. What function provides information into a smart device's internal diagnostics?

3. What function enables a device to output varying levels of its output signal?

4. Does a suppressed signal state that the device's output signal is consistently lower than expected or higher than expected?

5. When the sensor body is replaced, what procedure must be performed to allow accurate calibration of the device?

chapter 16

Fundamentals of Control Valve Maintenance

■ OUTLINE

■ OVERVIEW

This chapter provides information on the components that make up a control valve assembly. The components may not be physically mounted on or with each other, but all play a crucial role in the function of the control valve. This chapter gives the purpose for and the functions of all the appendages, and provides information on their working ranges and calibrations. This chapter does not cover valve breakdown and repair procedures—that is an in-depth area of study alone. It does provide information into the correct operation of all appendages and control valves.

■ OBJECTIVES

After completing this chapter, you should be able to:

- Explain the operation of a globe valve and a rotary valve.
- Describe the functions of a transducer, positioner, actuator, and regulator.
- Determine the proper operation of a pneumatic sliding stem and rotary actuator.
- Describe the operations of direct- and reverse-acting actuators.
- Explain the operation of a piston-type actuator.
- Discuss and explain the operation and calibration ranges of a pneumatic positioner.
- Identify and give the purpose for sliding-stem, guided, post-guided, and cage-guided valve assemblies.

■ INTRODUCTION

16.1 Control Valve Maintenance Concerns

We know from our previous studies that the objective of the control valve is to provide a method of controlling a process in a closed-loop design. A control valve is positioned so that a process flow is regulated to achieve a desired result. A control valve is not a device that can typically operate under its own guidance. An actuator is used to provide the necessary force to actuate the control valve. The actuator converts the signal that is received from the controller to a form that can be used to actuate the control valve (Figure 16–1).

A control valve may not be connected to an actuator. A valve can be manually positioned and, therefore, requires no automatic positioner. A valve may simply be an open/closed valve with no control position other than fully open or fully closed. This type of valve requires an actuator to provide the force and signal conversion to actuate the valve.

The pneumatic actuator is a proven performer in the "type" of actuator. This method of conversion of signal to force has been used for many years. We have studied previously how to calculate the working force of a pneumatic diaphragm. We know that the working force of supplied pneumatic pressure is used as the method of actuating a valve.

We have also studied how the valve diaphragm is opposed by a spring or coil of set tension. By applying a set pressure on one side of the diaphragm, the force becomes stronger than the opposing spring and valve stem movement occurs. By knowing the force opposing the diaphragm and knowing the supply pressure of the diaphragm, we can calculate a valve position. Sometimes it becomes necessary to verify and/or to require that a

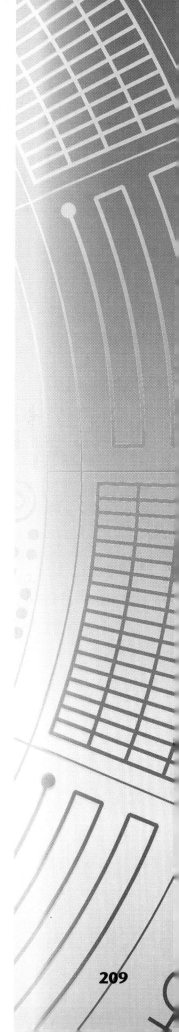

FIGURE 16-1 Fisher control valve with Taylor controller.

valve has reached a known position. This is where the use of a valve positioner comes into play.

A valve positioner is used for overcoming the forces that oppose valve movement. It is easy to picture a new valve that is installed into a process line and how easily that valve is actuated and positioned without any opposing forces other than those for which the valve is designed. Add the factors of time, process erosion, corrosion of moving components, and repetitive actions, and it becomes easy to see that original supplied pressure will not move the valve the same amount as originally installed.

A valve positioner is a repeat of the nozzle/flapper applications we studied earlier, but with a twist. Recall that the nozzle/flapper arrangement is a way to regulate an outlet pressure given a constant inlet pressure. The regulation component for a positioner is the valve position assembly. Linkages reflect a valve position on the positioner that is compared to the flapper position. If there is a difference, the nozzle is adjusted accordingly to place the valve in the correct position. A positioner is essentially a nozzle/flapper arrangement that acts as a proportional-only controller. The positioner acts when a difference is detected between the actual valve position and the desired valve position that is set by the controller.

If a difference, or error, exists between the actual valve position and the wanted valve position, the nozzle/flapper varies the air pressure that is applied to the diaphragm appropriately so the valve is pointed toward the desired position.

Let's make a list of the various interacting components of a control valve arrangement. We have the supply pressure to the valve diaphragm assembly, which has to be regulated by a (1) regulator. The (2) valve actuator is then used to force movement of the control valve stem, which forces control valve movement. The original position of the valve is taken care of by a (3) transducer (sometimes referred to as an I/P), which converts the controller's position signal (usually 4–20 mA) to a force used by the diaphragm (usually 3–15 PSI). Then we have a (4) positioner that can repeat or change the actuating pressure set by the transducer to reposition a valve.

Each of these components is susceptible to error. Therefore, it becomes necessary for a field technician to know how each component in the final control assembly interacts with the others. Also, a field technician must know how to maintain these components in good working order and how to repair them when they are not.

We have plenty to discuss with the above and we still have not made it to any actual (5) valve maintenance or repair. Actual valve maintenance and repair will be discussed after the above-mentioned because all of the components influence the control valve's performance. The components of a control valve assembly all perform the same function regardless of the manufacturer, so we can discuss them in general terms here. There is some review of material covered earlier to reinforce our working knowledge of each.

16.2 Pneumatic Regulator

A regulator is often used to maintain a constant working pressure (Figure 16–2). This piece of equipment is often thought of as being "secure," but that is not always the case. A regulator can be subjected to extreme temperatures and/or pressures that could damage internal workings. A regulator may vent the working gas to the atmosphere and care must be taken to ensure that the gases are not vented dangerously (towards personnel) or in a dangerous location (hazardous).

A regulator is often ordered and received with factory-calibrated working pressures. Most regulators have a point of adjustment of working (outlet) pressure. Care should be taken not to set the outlet pressure higher than the upper range limit specified on the regulator or damage can occur.

We see again the nozzle/flapper arrangement being used to directly control pneumatic pressure. By adjusting the flapper of the arrangement, the outlet pressure can be set to the desired pressure. The tension of the flapper can be adjusted to bring the working pressure back into tolerances specified for the valve assembly. Sometimes the valve assembly does not contain a regulator at the valve but, rather, has a regulator located at the point of the "instrument air supply." With such an arrangement, regulation is achieved the same way as at the valve but with a higher volume. Volume air flow is reduced for individual devices through the methods discussed in the "Fundamentals of Pneumatics" chapter—namely, boosters, amplifiers, and relays.

A regulator is generally mounted with or has the capability to mount a gauge on the downstream side of the regulator. Maintenance

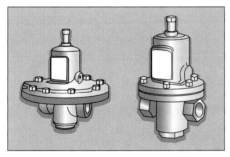

FIGURE 16–2 Pressure-reducing regulators.

procedures on regulators often consist of adjusting the working pressure by turning an adjustment screw on the regulator to the prescribed pressure indicated by this gauge. The adjustment screw of a regulator adjusts the tension of the flapper of the nozzle/flapper arrangement.

Adjusting pressures of a regulator requires proper ventilation of the pneumatic gas used. The vent port of a regulator needs to remain open at all times. The spring case vent hole should also remain open.

16.3 Valve Actuator

We studied the valve actuator (Figure 16–3) earlier in the "Fundamentals of Pneumatics" chapter. We know how to calculate the force of an actuator diaphragm's surface area from the applied pressure. We know that the control valve actually receives its commands from the actuator via the valve stem. A variety of components affect the method of how a control valve receives stimulation.

Maintenance procedures involving actuators vary and the individual device type guidelines should be referenced. Often, process industries have set procedures for valve actuator checkouts, repairs, rebuilds, calibrations, etc. Actuator parts, like any workable device, are subject to normal wear and tear. Actuators should be inspected visually whenever called for and when suspected of malfunction. The frequency of the inspection is often predetermined and is influenced by the type of service and the operating conditions of the actuator.

FIGURE 16–3 Direct-acting actuator.

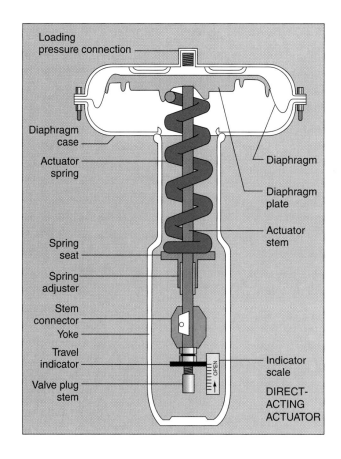

Most actuators can operate in either a vertical or horizontal position. Always check the manufacturer's specifications to eliminate a possible trouble spot.

Actuators do have calibrated, or adjusted, components that have to be checked on a regular basis: the spring adjustment, travel adjustment, and adjustable travel/stop.

The spring adjustment is an adjustable locknut (usually) that allows the actuator tension to be adjusted. This action can also affect its travel parameters. The adjustable/travel stop is related to the travel and spring adjustments and care should be taken when adjusting one of the parameters that the others are checked also.

Before you begin any work on the valve actuator, be sure to remove all sources of pneumatic and/or electric power. Be sure that the actuator cannot suddenly open or close. Only by blocking the process, supply power, and pneumatic pressure and following an established procedure can you ensure a safe work area and procedure.

Reducing the spring adjustment reduces the tension or seating force of a valve. If the valve is direct-acting, the spring tension varies the seating of the valve: the opposite applies if the valve is reverse-acting. If the spring tension is reduced on a direct-acting valve that is fail-closed, the valve seating, when closed, may not be enough to prevent leakage. If a valve is fail-open, the spring adjustment may prohibit the valve from reaching a fully open state if incorrectly adjusted.

Pneumatic valve actuators are generally either the diaphragm or the piston type (Figure 16–4). The diaphragm actuator is of one of two types: direct-acting or reverse-acting. Depending on the process operating parameters, the control valve will need to close or open. The direct- or reverse-acting actuators can provide the means for the movement.

The direct-acting actuator has its spring tension set to oppose the input pressure that positions the valve. The reverse-acting actuator has its spring tension set to reverse its forces to set the valve back to its original position. For either case, the components of the actuator will not vary but their mounting/connecting positions will.

A piston-type actuator is another type of valve actuator (Figure 16–5). A piston actuator will normally be used where the requirements of a diaphragm actuator did not have to be or could not be met. The piston actuator does not have the same force as the diaphragm but it is more compact. The piston actuator uses levers and angles of lever movement to transfer and increase its operating force. The combined use of a lever and operating angles allows a piston-type actuator to "build up" an adequate amount of operating force.

Due to its construction, a piston-type actuator can deliver force or torque action beyond what is available with a diaphragm-type actuator through its use of levers. A piston-type actuator is usually used under high operating pressure conditions and extra care must be taken to ensure that the actuator is "locked out" before any work is performed. A piston-type actuator is a good actuator for a quarter-turn valve, a valve that can open/close in 90° of travel (to be discussed later).

Regardless of the actuator type used, and the adjustment parameters for spring(s), tensions, and stops, all are adjustments that time and

FIGURE 16–4 Right angle valve actuator.

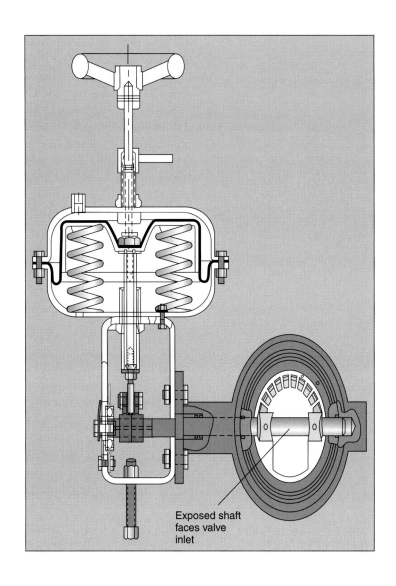

Exposed shaft faces valve inlet

FIGURE 16–5 Piston-type actuator.

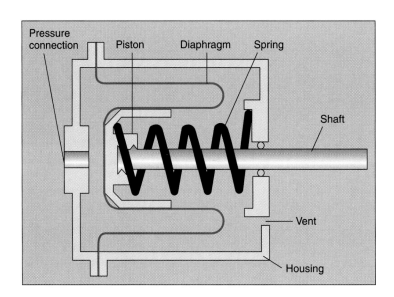

usage will change. True maintenance procedures should reference through experience which settings should be subject to the most deviation. The deviation from all the adjustments, in combination with friction and the nonlinear response of an actuator, can present a fairly common error in valve position known as hysteresis (Figure 16.6).

As with pneumatic instruments, we know that hysteresis occurs between the upscale and downscale valve positioning. The valve packing, nonlinear diaphragm construction, actuator adjustments, friction, and process pressure across the valve can contribute to the hysteresis until the level could reach as high as 10%.

We know that maintenance personnel are usually the ones who have to troubleshoot problems discovered in the control system. Troubleshooting a valve that does not position well in a process line can be the result of any one or more of the problems just mentioned. The actuator adjustments (in conjunction with positioner adjustments) can alleviate some of the error.

The nonlinear construction of the diaphragm can be explained by picturing a diaphragm with its opposing spring (Figure 16–3). As a small pressure (about 1 PSI) is applied to the valve diaphragm, the actuator will have a little movement because the spring tension is relatively small. As pressure is increased, the resulting diaphragm movement is made against a stiffer (greater) opposing tension.

Suppose the valve actuator has traveled 90% of its range. Apply 1 PSI of additional pressure to the diaphragm. It is easy to "picture" that the opposing force will be great enough to reduce actuator movement due to the increased tension of the actuator spring. This reduced movement

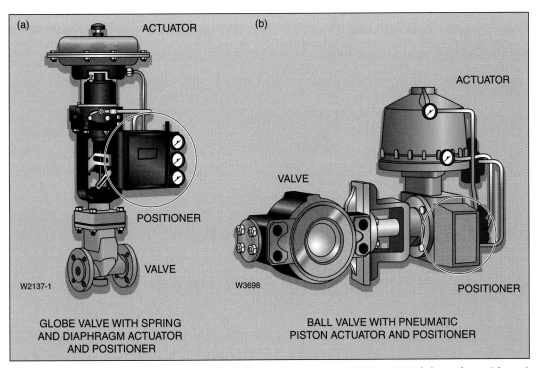

FIGURE 16–6 Globe (stem) and ball (rotary) valve assemblies. (a) Globe valve with spring and diaphragm actuator and positioner. (b) Ball valve with pneumatic piston actuator and positioner.

FIGURE 16–7 Fisher molded diaphragm.

compared to movement of the valve when the first PSI of pressure was applied gives the nonlinear relationship.

The valve does not travel the same amount with respect to the inlet pressure for its entire range. Why is this significant for actuator study? Well, as the diaphragm wears, it is more likely to become deformed or stretched (altered). To understand the consequences of a diaphragm's deformation we need to understand its construction.

Most pneumatic diaphragms are of the molded variety. Molded diaphragms are easier to install because the diaphragm is formed to fit exactly as it is installed. The molded diaphragm also provides a relatively constant working (surface) area for all amounts of valve travel.

By comparison, think of a diaphragm as a flat sheet. To cause any movement of the diaphragm due to inlet pressure, a **bubble** would have to be created to cause valve movement. This bubble opposes valve movement and is another contributor (albeit, a small one) to the nonlinear output of the valve position. This is not to say that in an emergency, a flat diaphragm could not be used to replace a molded one, because it can. A flat diaphragm should be replaced with its specific molded one as soon as possible (Figure 16–7).

To remove a diaphragm, first disconnect all process and pneumatic pressures and power supplies. There are tensions and pressures in an actuator that can cause severe harm. The process line should be blocked to prevent sudden process surges from repositioning the valve and, therefore, the actuator. The electrical signals could cause an unwanted spark, short, etc.

Some types of actuators do not require an external adjustment and, therefore, accessibility to the tension springs for release. Such actuators are generally used on rotary shaft valves and spring compression, or release for diaphragm replacement is not generally required.

The procedure for normal diaphragm replacement is a simple one. Remove the upper diaphragm case. If the valve is direct-acting, the valve diaphragm can be lifted out. If the actuator is reverse-acting, the head assembly (spring rods, tensioners, etc.) must be released to remove the diaphragm. When a diaphragm is replaced, the diaphragm casing must be uniformly sealed to prevent leakage.

Even if all the care in the world has been taken so that the actuator is performing as it should, a valve still may not be controlling as it is designed. A valve with hysteresis can make it seem as if the controller has too much proportional band and low gain, but the fault may lie with the valve assembly, such as the actuator. A positioner can be used to stop hysteresis errors, but first we need to understand how the signal is converted from the controller to a signal understood by the actuator.

16.4 Transducer

We became familiar with how a transducer operates in the "Fundamentals of Pneumatics" chapter. We should understand by now that the transducer is a method of converting energy of one form to another.

Most controllers today output a signal of the current type and most valves today use pneumatic pressure to force valve movement. The transducer converts the current (usually 4–20 mA) to a comparable

FIGURE 16–8 Current-to-pneumatic transducer.

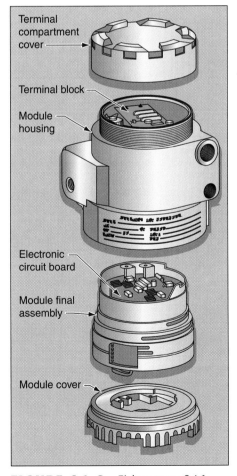

Terminal compartment cover

Terminal block

Module housing

Electronic circuit board

Module final assembly

Module cover

FIGURE 16–9 Fisher type 846 transducer mounted on actuator.

pneumatic pressure (usually 3–15 PSI). The use of the nozzle/flapper arrangement again provides the method of calibration.

Transducers also require calibration and the same characteristics of zero and span can be applied. Most transducers allow for field adjustments to be made even though the transducer was probably ordered with the correct working parameters for pneumatics and electric signals calibrated from the factory.

Remember, the working components of a device are subject to wear and deformation and, therefore, will need adjusting. The I/P will require that a specified current can be measured and the output pressure can be regulated by tuning the flapper to adjust the outlet pressure.

The same observations should be made of the transducer as made for other pneumatic devices. The proper venting of bleed pressures, the venting of exhaust pressures, and general working order should be verified. To be able to troubleshoot/repair a transducer requires that a field technician has an adequate understanding of how it operates.

A transducer is a very simple form of the nozzle/flapper arrangement. A coil is wound or fixed around a permanent magnet. When a current flows in the coil, the electromagnetic forces and the permanent magnetic forces are in direct opposition. The opposition of forces causes a flapper movement to occur, forcing the flapper to open or close depending upon the construction. A nozzle/flapper could be designed so that it opens or closes when current is applied.

A simple self-test using working pressures and currents can quickly be performed to determine an I/P's design. Connect a current supply to the transducer, run through the range of the input signal (usually 4–20 mA), and record the output pressures of the transducer. You should be able to determine if the transducer is direct-acting and follows 3–15PSI from 4–20 mA, or reverse-acting and follows 15–3 PSI from 4–20 mA. That is the crux of the maintenance problems of an I/P. If the construction type of the I/P is known, the I/P can be adjusted to output 3 PSI for 4 milliamps or 15 PSI for 4 millamps, or reverse-acting. Reversing a transducer is accomplished by switching the input leads and recalibrating for the desired range (Figure 16–8).

Most transducers include a booster (sometimes called a pneumatic relay) that allows for the boosting of the pneumatic pressure. These boosters are the same as the relays we have studied before, and we know that relays are capable of amplification. By using boosters, a transducer is available with a number of input and output ranges for various configuration properties such as split ranging.

Split-range systems are designed such that a control valve can perform from the 3–9 PSI range while another can operate from the 9–15 PSI range. Split-range designs provide a wider range of control that, perhaps, cannot be controlled by a single valve. The calibration of the transducer makes this possible. The versatility of the transducer is evident in its ease of calibration.

A transducer can be calibrated over a wide range including split-ranging, but it is also reversible. Again, working pressures are involved. You should use care when you disassemble transducers and when you perform other types of transducer maintenance procedures (Figure 16–9).

16.5 Positioners

In a closed-loop system, the goal is to keep one or more process variables within tolerances specified by the controller. We have read how the build-up of friction, hysteresis, and other factors can cause an incorrect positioning of the control valve. In a closed-loop system, there may be a small allowance or tolerance in the finished product. The small tolerances for error sometimes require the use of a positioner to "position" the valve as it is signaled to be placed.

The signal from a controller is interpreted by the transducer. The corresponding pressure is applied to the diaphragm and the valve stem changes position accordingly. We know the friction from the valve packing, the hysteresis from actuator response, the friction from the process flow, etc. can cause an incorrect positioning of the valve. When errors are introduced in the valve positioning, a positioner can be used to alleviate most of them.

A positioner is also used when accurate response is demanded quickly for a process variable. The positioner receives its input from the controller and determines from the controller's signal what the desired valve position should be. A feedback assembly, mounted on the valve stem, provides feedback to the positioner, which is used to compare the actual stem position to the desired position.

If corrections are needed, the positioner calls for an actuator adjustment until the desired and actual positions are the same. The comparison of positions is done mechanically and the output is a pneumatic pressure.

The pressure used by a positioner is often separate from the pressure that is used to force diaphragm movement. The different pressure sources allow for a positioner that is using the conventional pneumatic range of 3–15 PSI to control an alternate pressure, such as a 6–30 PSI range.

To calibrate a positioner we must recall the concepts of zero and span. A positioner is calibrated by the same method that we studied earlier, but before calibration starts, all of the relevant information has to be gathered. Signal pressure, full- or split-range, supply pressure, actuator working pressure range, valve travel, and whether the action is forward or reverse acting must be determined before the necessary calibration procedure can begin. The calibration of a positioner has to assume that the actuator is adjusted within its proper working range (Figure 16–10).

The objective with positioner calibration is to establish an accurate starting point referred to as zero. The controller is manipulated to send an equivalent zero signal (4 mA) and the positioner compares the *controller's* signal to the valve position. The positioner is adjusted such that at the zero signal, the positioner is calling for a zero position. As the signal is increased, the positioner should be adjusted (if needed) to call for an increase in the actuator working pressure.

Therefore, the two variables that need to be monitored are the controller's signal and the valve stem position. Adjust the zero setting to allow the stem position to remain at zero and, accordingly, throughout the valve stem range. If the valve has a linear response, the valve stem

FIGURE 16–10 Positioner characterization. Stem position feedback linkages often include a cam, as shown in part (a). The contours of the cam can be designed to provide different relationships between the input signal and valve travel. This ultimately results in a well-defined relationship between the input signal and control valve, C_v (flow). By selecting different cams, the system can be characterized according to special needs. For example, there may be a requirement to increase or decrease valve response at low lifts.

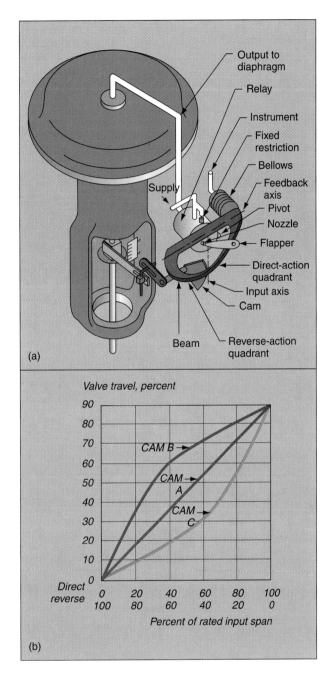

position should be at 50% for a 50% controller signal. The span adjustment is adjusted until the span of the valve corresponds with the controller signal. The valve stem should indicate full travel when 100% of controller signal is received.

When setting the span for valve positions, it is important that the span adjust is not made to where it takes an exaggerated signal to move the valve from its furthermost point of travel. Some manufacturers suggest that the valve travel for 100% should be set at a value slightly less than full range. Likewise, the valve should be adjusted fully closed when a signal slightly greater than zero position is received.

There are reverse-acting positioners and actuators; therefore, it is necessary to obtain the proper "specs" for each assembly before

FIGURE 16–11 Fisher control valve.

proceeding. When adjusting for span in the positioner adjustment, it is essential that zero is referenced again because the two do interact.

16.6 Control Valves

Control valves are used in industry to regulate a process by controlling the rate of flow or supply (Figure 16–11). By adjusting the opening in the valve assembly through which the controlled process flows, the valve regulates the **volume flow rate**. The operation of final control elements (control valves as well as other devices) includes the steps needed to convert the control signal from a controller into an appropriate response.

A control valve is often referred to as a *variable orifice* in the line. The equation studied earlier, "Bernoulli's theorem," can be used to calculate the differential pressure and resulting flow that passes through a valve. In short, the flow through a valve is proportional to the square root of the differential pressure across the valve times the area of the valve opening.

The difference with control valves is that the control valve variables, flow rate, and area of valve opening are not constant. With an upstream process supplied by other means, the flow rate available at a control valve is constant, but the flow rate downstream of the valve varies with the rate set by the control valve. The result is a pressure drop across the valve that is dependent on the valve opening. The closer the flow's rate is downstream of a valve to the rate available upstream, the lower the pressure drop across the valve. Also, the lower the pressure drop is across the valve, the closer the flow rate is to its maximum.

The capacity of a control valve must be sized so that it controls a process without varying the controlled flow ineffectively. Flow valve characteristics can be calculated by using an equation that solves for the C_v, which is the unit of the flow capability defined as the number of U.S. gallons of water per minute at ambient temperature that will flow through a fully open valve with a 1-PSI pressure drop across it. It is important that a control valve is properly sized for economic reasons and controllability.

If a valve is oversized, or is larger than needed, the valve will not have enough "resistance" to properly control the flow except when the valve is almost closed. It is apparent that this valve will allow the required flow to pass through, but it is larger than is needed and will cost more. If the valve is undersized or is smaller than needed, the required flow will not be able to pass through the valve and the valve must be replaced with the proper valve, which includes a material cost as well as a labor cost.

An undersized control valve will never pass the required flow capacity, so it is easy to see that the control range of the valve is reduced significantly. With most undersized control valves, the usable range of the valve does not exist and the valve will function as a "block" valve. This means that the valve will open fully when "signaled" to control the flow and close fully when "signaled" to close. The result is a valve that has no controllable action.

When a valve is oversized, it will attempt to regulate the flow by throttling the flow rate. This simply means that the valve will operate at a nearly closed position and the full range of the valve is not used. In addition, when a valve operates at a nearly closed position, high flow rates will be present, which can erode the valve seat. The C_v rate is the accepted way that industry selects control valves for use, but the equation can be used to determine if the flow rate can cause damage from excessive pressure drop or high rates of flow. The equation for flow rates varies with the type of fluid controlled:

$$\text{Liquids: } C_v = Q \sqrt{\frac{SG}{dp}}$$

$$\text{Steam and vapors: } C_v = \frac{w}{63.3} * \sqrt{\frac{v}{dp}}$$

$$\text{Gases: } C_v = \frac{Q}{1,360} * \sqrt{\frac{Tf * SG}{dp * P2}}$$

where:

Q or W	= flow rate (liquid gpm, gases scfh, vapors lb per hour)
SG	= specific gravity
Tf	= flowing temperature in degrees (Rankine [°F + 460])
dp	= pressure drop in PSI ($P1$–$P2$)
$P1$	= upstream pressure at the valve (inlet PSIA)
$P2$	= downstream pressure at the valve
v	= downstream specific volume (cubic feet per pound)

Engineers use these equations to determine the actual response of a control valve to its process environment, but the field technician who understands the concept of Cv can intuitively decide if a control valve is or is not functioning properly merely by interpreting the results from the Cv equation.

When a control valve is installed with a positioner, it displays two attributes that we can recognize. A control valve will operate under a linear relationship that implies that for 25% of signal, a valve stem travel will also equal 25%. A valve will be nonlinear, which implies that for 25% of signal, a valve stem travel may vary more or less than 25%. This "nonlinear" travel may follow an "equal percentage" travel rate, which implies that for 25% of signal, the flow rate will equal 25%. This does not mean valve stem travel will equal 25%. A "square-root" curve may also be followed. Proper documentation should be available and you should check it before you take any valve "off-line." Check its proper response against its recorded response (Figure 16–10). Often enough, the conditions of the valve's operating environment can dictate whether a valve will require additional maintenance other than the maintenance required for flow characteristics (Figure 16–12).

FIGURE 16–12 Fisher Baumann wafer control valve.

A control valve is designed to operate without friction. If friction is detected, it will most likely be found where the valve stem enters the valve packing. The valve stem often becomes corroded at the valve packing due to leaks in the packing itself and these leaks should be stopped. Excessive tightening of the valve packing can also cause friction. In this case, the valve packing is too tight against the valve stem, hindering the proper movement of the valve stem.

The air supply to a control valve should be checked for leaks, which will cause the valve to stroke improperly when signaled. The valve actuator will often leak where the diaphragm is "squeezed" between the two halves of the diaphragm case. In some valves, an "O-ring" is found where the valve stem leaves the diaphragm case; this "O-ring" can also leak. If the vent on a valve is plugged, improper operation may occur because added "back pressure" can hinder the valve. A less common valve leak, and one that is harder to locate, is a valve with a hole in the diaphragm. All valve leaks should be repaired when it is determined feasible to do so.

Inner valve wear can occur on all control valves that have been in operation. Corrosion, cavitation, erosion, pressure, and foreign material trapped in a valve are all causes of valve wear. These conditions will degrade the operation of a valve, sometimes to the point where they cannot be tolerated and must be corrected. The only way to properly check and repair a valve is to remove it from service and disassemble it.

If a control valve suddenly quits responding to an operator's signals, chances are it has lost the stem plug that allows the valve to be forced to its correct position. Remember, the valve stem is the part of the actuator that will control the movement of the valve. In this case, the valve will often swing to a position that is determined by the process.

If a valve does not stroke properly, you should check for the causes just listed. Ensure there are no air leaks and no obstructions in the valve body. Make sure the stem is controlling the valve body and there is not an excess of friction on the valve stem. If the valve still does not stroke properly, check to see if the valve actuator spring has lost its tension to control the valve. Also, the diaphragm can be stretched from excessive air pressure, to the point where it can no longer properly control the valve. If a diaphragm is worn out, the valve may attempt to function correctly at a certain pressure but not at any other pressure regardless of the tension of the spring adjustment.

For valves that require air to open and are vented inside the diaphragm casing with a vent located at the top of the diaphragm, moisture can build up in the casing and begin to erode the diaphragm, stem, casing, etc. In this case, remove the valve and turn it upside down. Supply air to the normal inlet to force the water out of the casing. In some cases, it may be necessary to disassemble the valve to dry the moisture points inside.

All valves should be equipped with "bug screens" to prevent blockages and foreign materials from entering the valve. If the valve strokes slowly but reaches its full range, chances are the vent is plugged.

Remember, when you disassemble a valve, follow the procedure established by the area where you work. Control valves in process lines often contain hazardous materials and should be decontaminated. Before you disassemble, remove any tension springs. Be careful of sharp edges caused by erosion.

For proper care and maintenance, it is important to follow the manufacturer's guidelines for the correct procedures. As a rule of thumb, plug valves are the only valves that require internal lubrication. Most valves require lubrication in the bushings located at the top of the yoke and on the stem threads, where exposed. If a valve is in a dusty environment or operates in a high-temperature area, it may be necessary to use a dry lubricant such as graphite to ensure proper lubrication. Some valves in a process line may be seldom, or even never, used. These types of valves are usually shutoff valves used only during shutdown or emergency procedures. Shutoff valves require maintenance to ensure that the valve will function when required. Maintenance of shutoff valves is usually limited to downtime and the maintenance is usually more thorough and determined by the specific site procedures for emergency shutdown I/O.

The aforementioned devices (1–5) 16.2–16.6 are usually the responsibility of the field technician to maintain in an efficient working order (Figure 16–13). All of these devices have set maintenance/repair procedures that are to be followed and minimum work and downtime of the process is the result. The difficult maintenance/repair procedures for valve bodies, and the fact that the valve body is the only segment of the control valve assembly that cannot be visually inspected, are what separates the valve body from the others. We are able to visually inspect the control valve assembly with all of the other appendages and could make a reasonable judgment as to whether the valve assembly is in adequate working condition.

FIGURE 16–13 Typical rotary valve actuator.

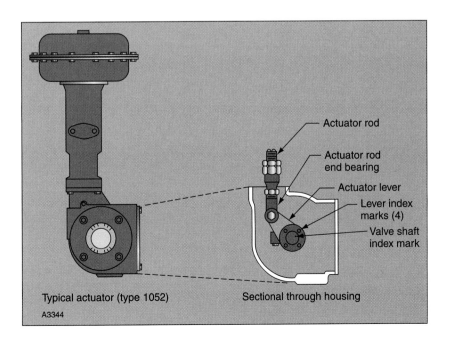

Our study so far in this chapter has been limited to the appendages of the control valve assembly and the workings of the valve body. All of the accessories attached to a control valve are present to provide an efficient method for the control valve to maintain position.

We know the control valve functions by regulating flow but now we will learn the basic type of valves and how they function. Valve bodies contain the process flow and the resulting process pressure drops across them. Valve bodies also must provide a means for the other appendages to be attached to the valve body.

Typically, the sliding stem globe valve is a standard valve for standard performance guidelines. A sliding stem valve is often referred to as a globe valve. The "globe" name comes from the chamber of the valve body through which the process flows.

The reason the sliding stem valve is not the only valve in use is that a globe valve often causes the process flow to change direction many times before it can exit the valve body. It is easy to picture a process flow of a high volume and capacity with resulting pressure drop, causing a significant amount of erosion in the valve body.

The result is there are several types of valves that allow for different services. The sliding stem (globe valve), rotary, and butterfly valves are probably the most common types. Each valve type has its own procedures for valve disassembly/assembly and repair. Procedures even vary among manufacturers for the same type of valve. The point is, the manufacturer's guidelines, along with any variations from the site where you work, should contain all necessary information to evaluate and repair the valve body.

We next discuss the general outlines for the predominately used valve types: sliding-stem and rotary (Figure 16–14).

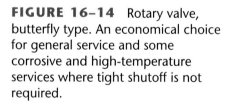

FIGURE 16–14 Rotary valve, butterfly type. An economical choice for general service and some corrosive and high-temperature services where tight shutoff is not required.

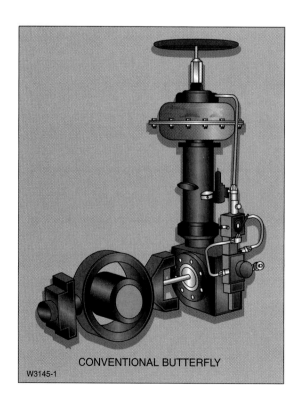

CONVENTIONAL BUTTERFLY

W3145-1

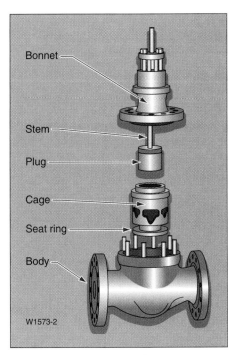

FIGURE 16–15   Sliding stem.

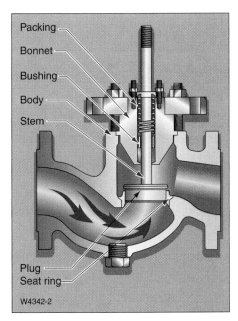

FIGURE 16–16   Sliding stem, stem-guided valve.

Sliding-Stem

The sliding-stem control valve has been the mainstay of the process control environment. This valve is not the only type of valve that is used for control, and is there is more than one type of sliding-stem valve. The sliding-stem valve has the conventional body, the slant body, and the angle body types. In addition, each type of sliding-stem valve has its own method of performing. Some are stem-guided, some are post-guided, and some are cage-guided (Figure 16–15).

Practically all stem- and post-guided valves follow the same convention. Stem- and post-guiding valves are relatively open and allow an easier flow through them compared to the cage valve. The slurry and the gritty-type stem- and post-guiding valves are good performers (Figure 16–16).

Stem-guiding types are typically used at a lower operating pressure than the post-guided type. The post-guided valve is a better performer in the high-pressure drop area and is better at reducing vibration, noise, and trim wear. The post-guided valve sometimes may have two posts to provide greater stability in higher-pressure drop process lines (Figure 16–17).

One good note about both types is that the valves are placed into the valve body through a top entry point. This makes it easy to inspect the valve and remove various components without removing it from the process line.

The limitations of a control valve may be the cause of maintenance work because the limitations are also the breaking point for components in the valve assembly.

There is a pressure point that will cause damage to a control valve. Sliding-stem and post-type valves generally work at the lower end of the pressure range. The valves tend to chatter at higher pressures, causing seat and trim damage. The valve may tend to slam open or closed depending upon the throttling action taken. The high pressures also create uneven forces on the valve seat and trim, which could cause increased damage to the valve chamber.

Valves are often checked by seating the valve and checking to see if a good seat was made. Dye, paint, and lapping compounds are some of the methods used to check valve seating. The trim and seat are compared after they are seated to see if the seating is even throughout the seating area called a seat ring. You can anticipate problems if you know that a post- or stem-guided valve is working in the upper ranges of its design with respect to pressure drops. Again, you need to understand how a device works before you perform a maintenance procedure.

Cage-guiding-type valves generally pick up where the sliding-stem and post designs leave off. The cage design allows for an even alignment throughout the entire range of the stroke of the sliding-stem. The stem- and post-guided valves tend to be pushed to one side when subjected to higher pressures and volumes but the cage assembly has the force spread out to cover the entire trim area of the valve, so valve seating is not endangered.

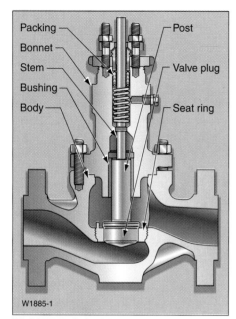

FIGURE 16–17 Sliding stem, post-guided valve.

The cage provides a method for rigidly seating the valve without the sideways push that is present to prevent chattering or noise. Another good principle of the cage valve, and a point that must be understood by the fieldhand, is the cage valve has different cages for different flow characteristics (Figure 16–18).

The cage openings through which flow takes place can be made to allow for certain types of flows. With a linear flow, the valve strokes allow an equal amount of flow to pass. When valve assembly is performed, the desired flow characteristics must be verified. A valve that should cause linear flow characteristics will not have the same cage assembly as a quick-opening type.

Rotary Valves

A rotary-type valve uses a quarter turn of the valve stem to position a valve closure member. There are three types of a rotary valves: partial ball, full ball, and disk.

A quick way to determine the actuation of a control valve is to visually inspect the valve assembly. A rotary-type valve will not have a stem that is used to "push or pull" the valve into a correct position. A rotary valve still has a stem, but it is used to position a valve by rotating the stem until the correct flow is passing through the valve.

A variety of rotary valve actuators are available. When adjusting the actuator, you must determine the characteristics of the actuator— is it push to open or push to close? The connecting linkage for the actuator to the valve body contains a lever that is connected to the stem by a key and keyway, or the lever is mounted on splines of the stem. Most actuators are equipped with lines showing the orientation of the stem/lever position.

FIGURE 16–18 Three cage designs for characterizing flow. (a) Quick-opening cages provide an inherent flow characteristic in which there is maximum flow coefficient with minimum travel. (b) Linear cages provide an inherent characteristic that can be represented ideally by a straight-line plot of flow coefficient versus percent of rated travel. (Equal increments of travel produce equal increments of change in flow at a constant pressure drop.) (c) Equal-percentage cages provide an inherent flow characteristic in which a given percentage change of valve will produce an equal percentage change in the existing flow coefficient.

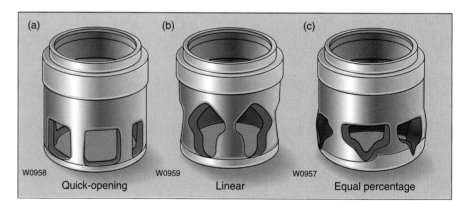

The predominant testing procedure for a rotary-type valve determines if the valve is fully opened or fully closed. To do this test, remove the valve from its process environment. Check the nose and trailing edge of a disc-type (butterfly) valve for clearances when open and closed.

A ball-type valve will also need to be removed for testing. The ball-type valve is a rotary-type valve that, when open, provides an easy path of flow for a process with a minimum pressure drop across the valve.

Butterfly valves get their name from the operation of the valve closing member. A circular disc is used to rotate it across the opening. Butterfly valves were typically used only for low-pressure processes. New valve linings and actuation methods are now allowing these types of valves to be used for high-pressure applications.

Butterfly valves still tend to slam closed when almost closed and to slam open when almost fully open. Fish-tail designs help to alleviate this tendency by diverting flow vanes and the use of the actuator. Butterfly actuators use a positive gear system that allows for the butterfly to be in the position only as the actuator allows. Despite the tendency of butterfly valves to slam shut or open, these valves remain some of the largest valves in operation—some valves are in excess of 20 feet in diameter. The simplistics of design, various designs, ease of maintenance, and the few moving parts are all reasons to use a butterfly valve when possible.

FIGURE 16–19 Fisher control valve with manual operation handle.

FIGURE 16–20 Inside a Fisher controller.

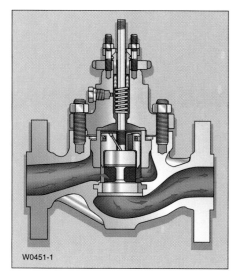

FIGURE 16–21 Single-port, sliding stem. Single-port valves have evolved as a modern standard for controlling a broad range of fluids under widely varying service conditions. Compared to double-port designs, they are more efficient, less costly, and easier to maintain.

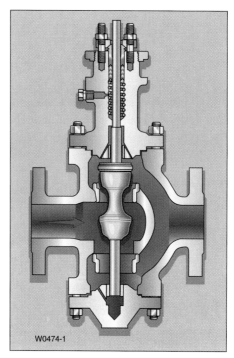

FIGURE 16–22 Double-port, sliding stem. Double-port globe valves similar to this early-pressure-balanced design remain in use today. Although the tortuous flow path may limit effective valve capacity, this configuration allows the passage of solids suspended in the flow stream.

FIGURE 16–23 Sliding-stem, balanced-cage type valve.

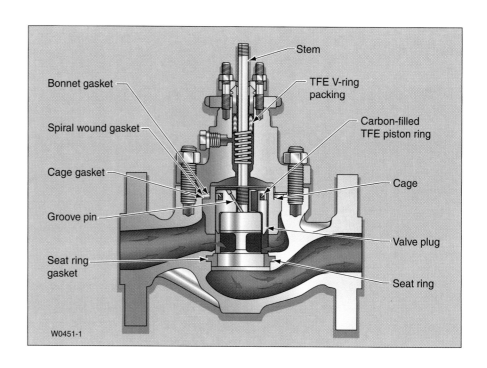

Stem

Bonnet gasket

TFE V-ring packing

Spiral wound gasket

Carbon-filled TFE piston ring

Cage gasket

Cage

Groove pin

Valve plug

Seat ring gasket

Seat ring

FIGURE 16–24 Alternate body styles. Globe valves present a tortuous flow path when compared to slant or angle designs. A general rule for predicting relative valve capacity is that fewer turns in the flow path allow higher capacity for a given valve size. For example, while a common single-port globe valve (having four flow turns) supplies 16.5 C_v per square inch of port area, an angle body (with only one flow turn) may have nearly twice the capacity per square inch of port area. Slant valve design capacity falls between the two.

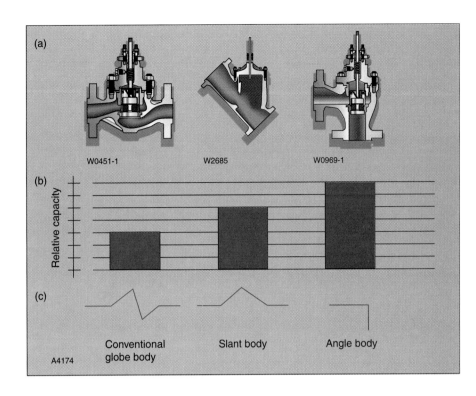

FIGURE 16–25 Alternate body styles. (a) Angle bodies provide a self-draining feature. They also fill the need for a control valve that is also an elbow. (b) Push-down-to-open valves incorporate a bottom flange to allow access to the trim and other internal parts. (c) Three-way cage-guided valves are used in a variety of flow mixing or diverting services.

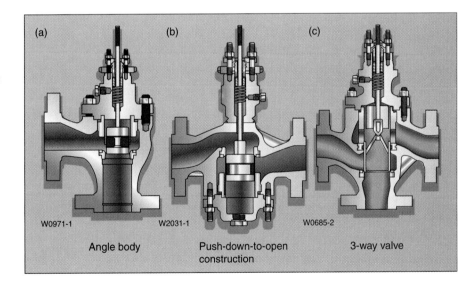

■ SUMMARY

This chapter provided the information for analyzing a field control valve for direct or reverse acting, actuator type, positioner calibration, and function and transducer operation. This chapter also explained the function of the control valve assembly and described its components.

■ REVIEW QUESTIONS

1. What opposes the force of pneumatic pressure applied to the control valve and attempts to return the valve to its normal position?

2. What is the function of a pneumatic regulator?

3. Describe the actions of direct-acting and reverse-acting actuators.

4. A piston-type and a rotary-type actuator achieve full valve movement in how many degrees of turn?

5. What is the function of a transducer?

6. Give two reasons why a positioner may be required.

7. What are the two different responses of the control valve, when coupled with a positioner, given an input signal?

8. Which type of valve body assemblies is designed to operate under higher pressures—the sliding stem or the post-guided sliding stem?

9. Why would a gauge-guided-type valve assembly be required?

10. A ball valve and a rotary valve operate with what type of actuator?

Fundamentals of Instrument Tubing

■ OUTLINE

■ OVERVIEW

This chapter provides information that is often overlooked by other references that cover instrumentation. This chapter provides the necessary information to explain why process tubing is used. The proper measurement applications for gases and liquids are discussed along with the effects of using filled reference legs, often called *wet legs.* Bending, marking, and supporting information is provided as well.

■ OBJECTIVES

After completing this chapter, you should be able to:

- Explain the purpose of using impulse tubing (process tubing) to obtain a process measurement.
- Describe the application of wet-leg process tubing.
- Explain the results on calibration when using filled reference legs (wet legs).
- Describe and perform correct installations of process tubing.

■ INTRODUCTION

17.1 Fundamentals of Instrument Tubing

Instrument tubing, sometimes called process tubing or impulse tubing, carries a process variable pressure to the sensor. A process change is reflected through the tubing to a sensor, which measures these "impulses" as a process variable.

By now you have probably figured out that practically all instrumentation that performs one of the three major process variables (pressure, level, and flow) does so through the measurement of pressure—more importantly, through the pressure differential. The fourth variable, temperature, can also be measured from a pressure sensor.

Impulse tubing is the tubing run that connects the transmitter sensor to the point of process measurement (Figure 17–1). Process tubing, therefore, provides the path for the process variable to reach the measurement's sensor. An incorrect path, mounting, or support for impulse tubing can generate an error in the measurement of the process variable just as surely as can a faulty or miscalibrated transmitter.

17.2 Purpose

Improperly run impulse tubing may be a more significant error than a miscalibrated transmitter. Suppose tubing was run for a differential pressure transmitter that allowed air to accumulate in the tubing at the transmitter. The tubing did not have a method for gaseous detection such as a "drain," nor was it run to allow drainage at the transmitter. The partial change equal to the amount of "head" sensed by the high or low side of the differential pressure transmitter will be reflected by the transmitter's signal.

There are times when a transmitter shows a sudden change in a process variable. Such changes, such as tap plugging, a crimped impulse tubing that becomes plugged, etc., are easily detectable. A rapid change presented to the control system will almost always be noticed by the controlling equipment, an operator, or the worker responsible for the instrumentation.

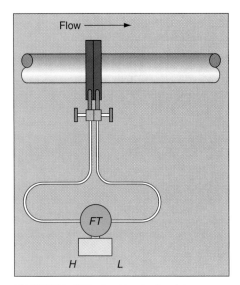

FIGURE 17–1 Typical tubing installation with the use of isolation or lock valves located at the process taps. A three-valve manifold for transmitter isolation is sometimes required for higher working pressures.

An impulse tubing line that allows moisture to build up slowly will present to the system an error that is slow to increase and, therefore, slow to detect. A slow change in the signal of a process variable will, perhaps, go unnoticed until the boundaries for a safe operating process are approached. The change is simply not noticeable when a technician may be watching several hundred I/Os that are active and changing with the process.

Often a moisture build-up in impulse tubing does not add or change a process variable but rather makes the measurement of the process variable sluggish. The moisture in the tubing acts as a weight that must be moved to allow the impulse to travel the length of the tubing. This creates a transmitter signal that varies as a pendulum does— initial movements follow measurements that appear to swing back and forth around the actual process variable.

17.3 Wet-Leg Tubing

Some impulse tubing is designed for fill fluids to be present in the "legs" of the impulse tubing. This impulse tubing is sometimes referred to as *wet-leg tubing*, or filled-leg tubing. Filled-leg or wet-leg tubing contains fluids in the process tubing that are used to transmit the signal of differential pressure. Sometimes the fluid in the legs of the transmitter tubing is not of the same type as the process fluid being measured. Special diaphragms are used to isolate the two fluids from coming into contact, but are still able to transmit impulses of pressure differentials.

Wet-leg tubing must be installed to allow the bleeding of any gases that become trapped in the tubing. Trapped gases in the tubing tend to prevent small changes in the process variable from being propagated along the length of the tubing. Large changes in the process variable may be delayed from being sensed by the transmitter. The trapped gas acts as a cushion in the tubing, absorbing process deviations until an impulse is strong enough to compress the trapped gas, causing the liquid on the other side of the trapped gas to be moved. The result is a process deviation being measured (Figure 17–2).

Wet-leg tubing demands that the legs of the high and low sides of the transmitter must be the correct length. If you are measuring flow, the legs must be the same length. If you are measuring level, you need to verify instrument mounting location and, subsequently, the impulse tubing, to avoid suppressing or elevating the signal.

17.4 Installation

Some processes call for special tubing. Corrosive chemical reactions may necessitate the installation of certain alloy tubing for certain processes. The specifications for the device installed (the spec sheet) should reflect the installation of special alloys, if called for.

A tubing installation that is installed correctly will have accurate bends resulting from accurate measurements. All tubing installations should be done in a neat and professional manner. Tubing is run precisely into fittings and the entry to tubing terminations must be

FIGURE 17-2 Wet-leg calibration range calculation.

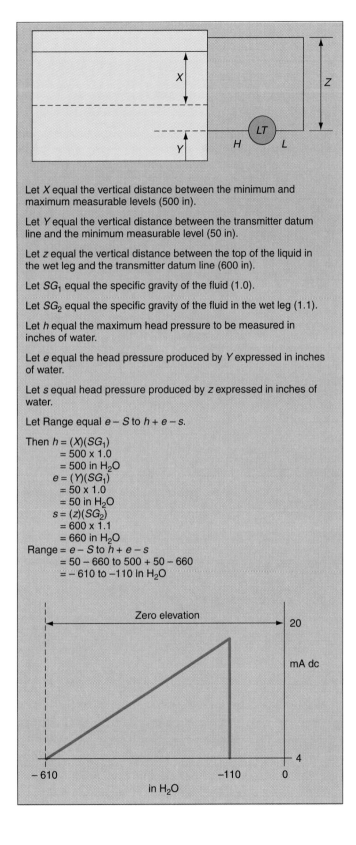

Let X equal the vertical distance between the minimum and maximum measurable levels (500 in).

Let Y equal the vertical distance between the transmitter datum line and the minimum measurable level (50 in).

Let z equal the vertical distance between the top of the liquid in the wet leg and the transmitter datum line (600 in).

Let SG_1 equal the specific gravity of the fluid (1.0).

Let SG_2 equal the specific gravity of the fluid in the wet leg (1.1).

Let h equal the maximum head pressure to be measured in inches of water.

Let e equal the head pressure produced by Y expressed in inches of water.

Let s equal head pressure produced by z expressed in inches of water.

Let Range equal $e - S$ to $h + e - s$.

Then $h = (X)(SG_1)$
$= 500 \times 1.0$
$= 500$ in H_2O
$e = (Y)(SG_1)$
$= 50 \times 1.0$
$= 50$ in H_2O
$s = (z)(SG_2)$
$= 600 \times 1.1$
$= 660$ in H_2O
Range $= e - S$ to $h + e - s$
$= 50 - 660$ to $500 + 50 - 660$
$= -610$ to -110 In H_2O

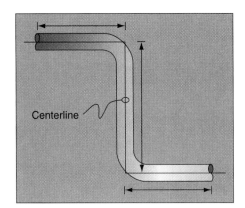

FIGURE 17–3 When measuring impulse tubing, measure from centerline to centerline of the bends.

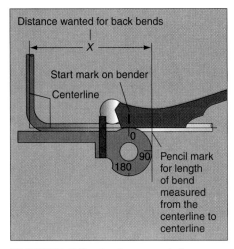

FIGURE 17–4 Start mark for back-to-back 90° bends.

straight. Installers should be aware that they are responsible for the tubing run they are installing.

When you measure for a tubing bend, measure along the centerline of the tubing. Tubing benders have a mark at the centerline of the tubing that signifies the edge of radius (Figure 17–3).

When you take your measurement, use a sharp pencil to indicate the measurement point. It is sometimes necessary to use a ferrule to mark around the tubing in a precise plane. Never use a sharp, penetrating object to mark on tubing. The scratches left on tubing can be a source of corrosion and a possible path for process fluids to leak.

A good rule to follow is to bend your tubing in only one direction. Since elevation changes in process tubing are significant, be sure to scribe a mark on the first piece of tubing installed for the top side and continue in one direction from there. By using scribe marks on your tubing, it is easier to bend a piece of tubing multiple times without having to put it in place each time to see how it fits.

Suppose you have two 90° bends that are used to change elevation. You would make your first bend and then measure for the second bend from the centerline to the centerline of the tubing. Place your bender in the same direction as the first bend with the top side reflected on the tubing, and bend until the 90° mark is reached (Figure 17–4).

As with bending conduit, there is a certain amount of spring-back that comes when bending tubing. A good rule of thumb is to use 2° to 3° of spring-back per 90° of bend. Remember that the bend angle will be magnified over great distances, so be sure to reach the correct bend before you install the piece of tubing. It is always best to under-bend the tubing first and then incrementally add angle to the bend so you do not go over the desired angle. It is hard to remove angle of bend from a piece of tubing without kinking or damaging the tubing.

Sometimes you may want to pre-measure several bends in the tubing run before you make the first bend. Remember, gain will be introduced into the tubing length. We usually measure out tubing runs on 90° angles but we do not allow for the radius of the bend not traveling the full length of the prescribed route. The result is gain in the instrument tubing and subsequent bends. A good rule of thumb is to allow one tubing diameter for each 90° bend made.

Once you have routed the tubing and placed it into position, you must clamp the tubing to stabilize it. Vibration will cause the tubing to loosen its fittings, causing fatigue to the point where it may rupture. Sags or crowns may appear to trap moisture or gases. Stress may be placed on fittings that will eventually break.

Currently, no nationwide standards apply to this industry. Individual locations may or may not have their own standards for impulse tubing mounting. The following guidelines are general rules of thumb.

1. Stack tubing to save space.
2. The distance between clamps for ¼-inch through ½-inch tubing is 3 feet.
3. For ⅜-inch through ⅞-inch tubing, the supports should be placed 4 feet apart.

4. 1-inch tubing should be supported every 5 feet and 1-¼-inch tubing should be supported every 7 feet. These are general distances given for a typical installation.

5. The presence of vibration, heavy traffic, etc. may call for the use of additional supports.

Tubing that is run to allow for no process variable errors is considered to be an *adequate* instrument run. A process tubing run that has no errors and is installed in a manner that prevents instrument errors from occurring and is neat and professional in manner is considered to be *preferred.*

FIGURE 17–5 Three-valve manifold tubing connections and isolation sequence.

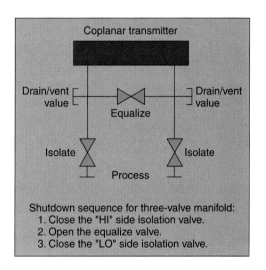

FIGURE 17–6 Process tubing and three-valve manifold.

■ SUMMARY

This chapter provided the basic information needed to route, support, and apply process tubing. Individual locations will have specific requirements for the types of tubing and fittings used. This chapter does not try to specify those requirements. It does, however, provide general information into the requirements for use and installation of process tubing.

■ REVIEW QUESTIONS

1. Give a brief explanation for impulse (process) tubing.
2. What consideration for process tubing routing must be considered if gases are to be measured?
3. Measuring and bending process tubing should always be referenced to where?
4. How far apart should supports be placed for ¾-inch tubing?

chapter 18

Fundamentals of Documentation

■ OUTLINE

■ OVERVIEW

Engineering drawings are the only link between someone who has created a control system and the person performing the actual installation. This chapter describes the different drawings available to technicians to help them accurately install a control system. This chapter provides a definition of and the purpose for each drawing as well as each drawing's relationship to the overall installation process. This chapter provides the necessary information to ensure that all documentation is explained and to ensure that all drawings are available for use.

■ OBJECTIVES

After completing this chapter, you should be able to:

- Describe the different drawings available for instrument installation.
- Determine the revision number of the drawings provided.
- Use "scope of work" changes on drawings to determine actual work identified.
- Use installation details to determine accurate installation requirements.
- Use P&IDs and other drawings to determine the function of the device(s) as required.
- Fill out a calibration data record properly and completely when performing a calibration procedure.

■ INTRODUCTION

18.1 Fundamentals of Documentation

The ability to read and understand information contained on drawings is essential to perform most engineering-related jobs. Engineering drawings are the industry's means of communicating detailed and accurate information on how to fabricate, assemble, troubleshoot, repair, and operate a piece of equipment or a system. To understand how to "read" a drawing, it is necessary to be familiar with the standard conventions, rules, and basic symbols used on the various types of drawings. But before you learn how to read the actual "drawing," you must have an understanding of the information contained in the various nondrawing areas of a print. This chapter will address the information most commonly seen in the nondrawing areas of a nuclear-grade, engineering-type drawing. Because of the extreme variations in format, location of information, and types of information presented on drawings from vendor to vendor and from site to site, all drawings will not necessarily contain the following information or follow the same format, but will usually be similar in nature.

18.2 Engineering Drawings

A generic engineering drawing can be divided into the following five major areas or parts:

1. Title block
2. Grid system
3. Revision block

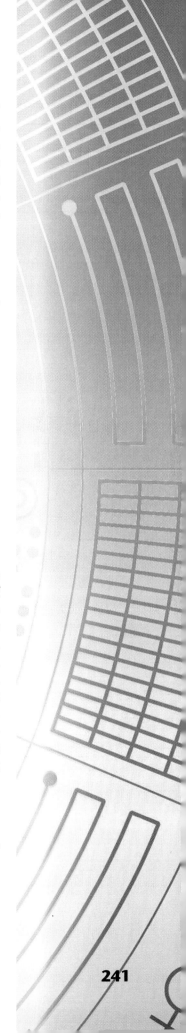

4. Notes and legend
5. Engineering drawing (graphic portion)

The information contained in the drawing itself will be covered in subsequent modules. We first need to cover the nondrawing portions of a print. The first four parts listed previously provide important information about the actual drawing. The ability to understand the information contained in these areas is as important as being able to read the drawing itself. Failure to understand these areas can result in improper use or misinterpretation of the drawing.

The Title Block

The title block of a drawing, usually located on the bottom or lower right-hand corner, contains all the information necessary to identify the drawing and to verify its validity. A title block is divided into several areas, as illustrated in Figure 18–1.

First area of the title block The first area of the title block contains the drawing title and the drawing number, and lists the location, the site, or the vendor. The drawing title and the drawing number are

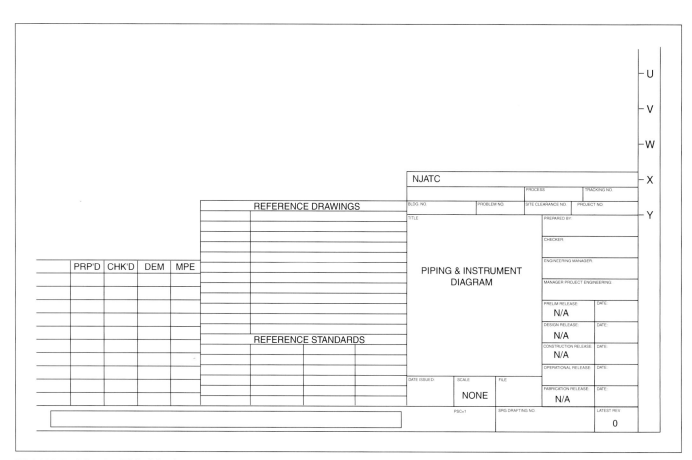

FIGURE 18–1 Title block.

used for identification and filing purposes. Usually, the number is unique to the drawing and is comprised of a code that contains information about the drawing such as the site, system, and type of drawing. The drawing number may also contain information such as the sheet number, if the drawing is part of a series, or the revision level. Drawings are usually filed by their drawing number because the drawing title may be common to several prints or a series of prints.

Second area of the title block   The second area of the title block contains the signatures and approval dates, which provide information as to when and by whom the component/system was designed and when and by whom the drawing was drafted and verified for final approval. This information can be invaluable in locating further data on the system/component design or operation. These names can also help in the resolution of a discrepancy between the drawing and another source of information.

Third area of the title block   The third area of the title block, the reference block, lists other drawings that are related to the system/component, or it can list all the other drawings that are cross-referenced on the drawing, depending on the site's or vendor's conventions. The reference block can be extremely helpful in tracing down additional information on the system or component. Other information may also be contained in the title block and will vary from site to site and from vendor to vendor. Some examples are contract numbers and drawing scale.

Drawing scale   All drawings can be classified as either drawings with scale or those not drawn to scale (NTS). Drawings without a scale usually are intended to present only functional information about the component or system. Prints drawn to scale allow the figures to be rendered accurately and precisely. Scale drawings also allow components and systems that are too large to be drawn full size to be drawn in a more convenient and easy to read size. The opposite is also true. A very small component can be scaled up, or enlarged, so that its details can be seen when drawn on paper. Scale drawings usually present the information used to fabricate or construct a component or system. If a drawing is drawn to scale, it can be used to obtain information such as physical dimensions, tolerances, and materials that allow the fabrication or construction of the component or system. Every dimension of a component or system does not have to be stated in writing on the drawing because the user can actually measure the distance (e.g., the length of a part) from the drawing and divide or multiply by the stated scale to obtain the correct measurements. The scale of a drawing is usually presented as a ratio and is read as illustrated in the following examples:

- $1'' = 1''$ Read as 1 inch (on the drawing) equals 1 inch (on the actual component or system). This can also be stated as FULL SIZE in the scale block of the drawing. The measured distance on the drawing is the actual distance or size of the component.

- 3/8″ = 1′ Read as 3/8 inch (on the drawing) equals 1 foot (on the actual component or system). This is called 3/8 scale. For example, if a component part measures 6/8 inch on the drawing, the actual component measures 2 feet.

- 1/2″ = 1′ Read as 1/2 inch (on the drawing) equals 1 foot (on the actual component or system). This is called 1/2 scale. For example, a component part measures 1-1/2 inches on the drawing the actual component measures 3 feet.

Grid System

Because drawings tend to be large and complex, finding a specific point or piece of equipment on a drawing can be quite difficult. This is especially true when one wire or pipe run is continued on a second drawing. To help locate a specific point on a referenced print, most drawings, especially piping and instrument drawings (P&IDs) and electrical schematic drawings, have a grid system. The grid can consist of letters, numbers, or both, that run horizontally and vertically around the drawing, as illustrated in Figure 18–2. Like a city map, the drawing is divided into smaller blocks, each having a unique two-letter or

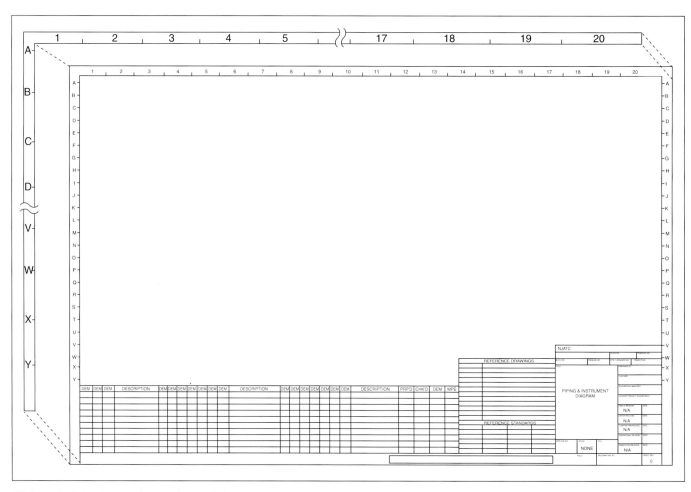

FIGURE 18–2 Grid-coordinated drawing.

number identifier. For example, when a pipe is continued from one drawing to another, not only is the second drawing referenced on the first drawing, but so are the grid coordinates locating the continued pipe. Therefore, the search for the pipe contained in the block is much easier than searching the whole drawing.

Revision Block

As changes to a component or system are made, the drawings depicting the component or system must be redrafted and reissued. When a drawing is first issued, it is called *revision zero,* and the revision block is empty. As each revision is made to the drawing, an entry is placed in the revision block. This entry will provide the revision number, a title or summary of the revision, and the date of the revision. The revision number may also appear at the end of the drawing number or in its own separate block, as shown in Figure 18–3. As the component or system is modified, and the drawing is updated to reflect the changes, the revision number is increased by one, and the revision number in the revision block is changed to indicate the new revision number. For example, if a revision 2 drawing is modified, the new drawing showing

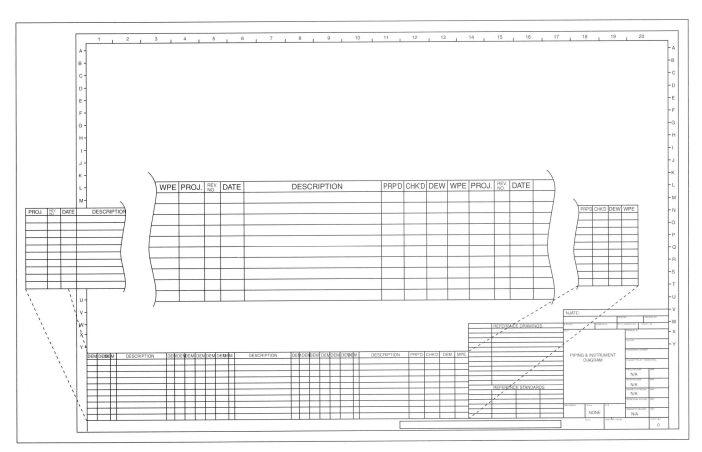

FIGURE 18–3 Revision block.

the latest modifications will have the same drawing number, but its revision level will be increased to 3. The old revision 2 drawing will be filed and maintained in the filing system for historical purposes.

Changes There are two common methods of indicating where a revision has changed on a drawing that contains a system diagram. The first is the cloud method, where each change is enclosed by a hand-drawn cloud shape, as shown in Figure 18–4. The second method involves placing a circle (or triangle or other shape) with the revision number next to each affected portion of the drawing, as shown in Figure 18–4. The cloud method indicates changes from the most recent revision only, whereas the second method indicates all revisions to the drawing because all of the previous revision circles remain on the drawing. The revision number and revision block are especially useful in researching the evolution of a specific system or component through the comparison of the various revisions.

Notes and Legend

Drawings are comprised of symbols and lines that represent components or systems. Although a majority of the symbols and lines are self-

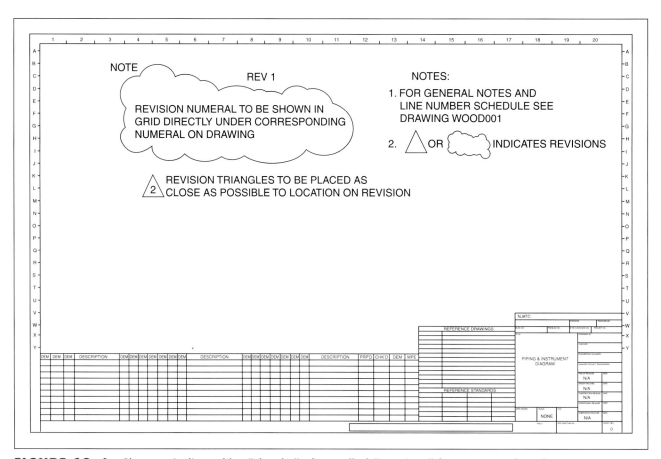

FIGURE 18–4 Changes indicated by "clouds;" often called "scoping," from scope of work.

explanatory or standard (as described in later modules), a few unique symbols and conventions must be explained for each drawing. The notes and legends section of a drawing lists and explains any special symbols and conventions used on the drawing, as illustrated in Figure 18–5. Also listed in the notes section is any information the designer or drafter felt was necessary to correctly use or understand the drawing. Because of the importance of understanding all of the symbols and conventions used on a drawing, technicians must review the notes and legends section before they read a drawing.

The documentation used by a field technician is considered the major link between the design process and the building process of an instrument installation. Field technicians will probably not work for the engineering design firm that has designed the installation, but they will use the firm's documentation to interpret the design.

It does not matter if technicians work in the construction or maintenance field; they must know the proper documentation for the job.

FIGURE 18–5 Notes and legend.

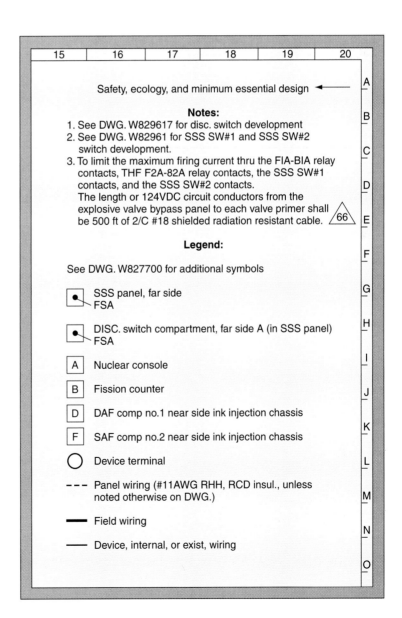

The very nature of instrumentation and/or control systems requires that time must be spent reviewing P&IDs, spec sheets, wiring diagrams, wiring installation details, etc. A significant amount of time on every project involving control systems is spent studying or reviewing documentation.

Although every job does not require every drawing in order to be completed, the majority will require the following: installation details, instrument index sheets, instrument specification sheets, loop sheets, panel drawings, plot plans, and P&IDs. There are a considerable number of other drawings that could be required for special-purpose jobs, but for general installation jobs these types are adequate.

18.3 Installation Details

Installation details show the intended design application of piping, tubing, material types (stainless steel, aluminum, etc.), electrical connections, wire size and number of conductors, conduit size, and other reference specifications as deemed necessary by the design firm. Installation details are usually job specific, which means that for an instrument mounting installation detail, the wire size, number of conductors, etc. will not be included. What will be included is the necessary instrument, mounting location, mounting height, material types to be used, bolt patterns, and any other information that may be needed.

When installation details are available and used properly, they can save many labor hours and will ensure the proper installation of the device.

For example, a differential pressure transmitter is to be used to measure the level of a HF acid tank. Installation details should show or reference all necessary information for the installation. Tap locations, device type, impulse tubing size, Monel construction/material type, mounting height to prevent elevation or suppression errors, conduit entry point, wiring terminations, Tag Number, safety considerations, and others that are deemed necessary are all required to install one device.

Installation details are intended to save time and to prevent the incorrect installation of a device.

18.4 Instrument Index Sheet

Instrument indexes provide a method for listing every instrument that will be used for the job. A summary that gives the present standing/location of the device usually accompanies each listing. An instrument index provides an accurate way to gauge the amount of work that must be done or remains to be done.

An index should be started at the beginning of the job and, as items are added, the list increases. Relevant shipping and receiving dates should be included. The changes that are made to the job, and how they affect the instrument index, should also be reflected. A device once entered into the list should never be removed—just updated to show present status.

An instrument index should contain the devices that are to be included in the present job requirement. This is a minimum requirement—each job will require its own variations.

18.5 Instrument Specifications Sheet

Instrument specification sheets or, more commonly, spec sheets, are used to tell a variety of people a wealth of information about a device. Spec sheets are generally sorted by Tag Number and are usually located in a permanent records library.

Spec sheets serve as a permanent record of devices that are in use or have been in use. The process conditions, calibration ranges, Tag Number(s), transmitter type, control functions, and others are included on the spec sheet. Engineering considers spec sheets to be the primary source of instrument information when ordering, designing, controlling, etc., a process.

Purchasing departments often refer to the spec sheets to determine the number of devices on hand compared to the average failure rate to get a rough idea on the estimated repair costs. Spec sheets provide an easy path for interaction between the engineering staff and front office personnel.

Spec sheets will also contain the necessary information for field personnel to maintain and install the instrument. Sometimes the spec sheet is not completed until after installation to allow for the checking and reviewing of a device. For such devices, the instrument index sheet will be the guiding documentation until a spec sheet is completed.

18.6 Loop Sheets

Loop sheets are arguably the document that will receive the most use by installers, troubleshooters, and maintenance personnel. An instrument loop sheet is as its name implies—a loop sheet that depicts the actual schematic wiring diagram of the field wiring.

The current path on a loop sheet can be traced from beginning to end until it traverses a complete loop. Each device that will function within that wiring loop is depicted. The simplest loop sheet is one that shows only a transmitter and controller, and nothing else.

More complicated loop sheets will have recorders, alarms, pilot lights, gauges, control valves, etc. The common point to a single loop sheet is that the devices on each sheet are generally categorized by instrument Tag Number. Recall that the Tag Number is used as an identifying method for records, accountants, material handlers, and for the field technician. By having items with the same Tag Number shown on the same loop sheet, a simple method of data keeping is available to all involved.

For the field worker, the loop sheet shows what devices are likely to be affected, where the device is located, the power requirements of that device, the controlling method for that device, the intermediate wiring, junction boxes, terminal strips, and the list goes on.

Jobs that have pneumatic instrumentation involved also have a loop sheet. Instruments with Tag Numbers are shown along with their hook-up details.

Loop sheets are printed in a variety of sizes, but most are printed on 11 × 17 (B) size drawing paper. The information that is to be displayed often is the determining factor for paper size. At some locations, a standard loop sheet size will be adopted and all documentation is expected to follow that size.

The loop sheet is the field worker's link to the process. It is conceivable for all as-built drawings to be submitted in loop sheet form for an instrument job. The as-built drawings can be subjected to a reverse design procedure that can fill in the gaps where associated drawings should indicate new work or changes in old work.

The loop sheet is also used during start-up and loop-checking procedures. Loop-checking procedures follow the path of wiring that is indicated by a loop sheet until all possible error points are confirmed as being error free.

The simple example of a loop sheet with a transmitter and a controller can be checked by forcing a signal on the pair of wires that are connected to a transmitter and then monitoring the response of the controller to verify that the associated wiring is correctly installed along with the controller's configuration limits.

In other words, a 4–20 mA signal is applied to the pair of wires to the controller and then the signal strength is adjusted. The signal will be read in by the controller and it should be compared to the sending signal strength to verify proper working order.

A loop sheet often lists calibration data such as calibration range and any associated setpoints. Field workers must be aware of the hierarchy of the drawing system at the site where they are working. For most locations, a specification sheet will list what is considered "superior" data. Therefore, if a loop sheet specifies a calibration range and the spec sheet does not agree, the spec sheet will be taken as being correct. If a conflict exists between two forms of documentation, the worker is to call attention to it before doing any work.

A loop sheet may list incorrect data pertaining to calibration ranges, setpoints, etc., and field workers must be aware of possible conflicts before they start the job. There have been numerous accounts of industrial shutdowns due to incorrect data on loop sheets that was taken for granted to be correct when the hierarchy of the documentation system stated that the specification sheets had precedence over loop sheets.

It is easy to see that if the only drawings that are used for as-builts are the loop sheets, technicians must take extra care to ensure that the proper documentation updates are carried out.

18.7 Panel Drawings

Panel drawings are generally a simpler approach to layouts of monitoring, recording, and transmitting equipment. Panels are mounting points of availability for devices to be mounted on so that the devices are accessible and observable to operators and/or instrument workers. An instrument panel can have a variety of forms and concepts but the panel drawings generally are the same.

Panel drawings show, first of all, where devices that are mounted on the panel are located. The devices are given locations along lines to allow for ease of installation and for maintenance procedures that will come later.

An "elevation" view is one form of information that is contained on a panel drawing. Another form is to show what is behind the panel.

The interior wiring of all the devices that are mounted on the panel is often shown on a sequence of drawings that are called *panel drawings.* Like other drawings, a hierarchy must be established in the area where you work. There will be conflicts among panel drawings and loop sheets, wiring diagrams, plot plans, and even P&IDs. The important thing is to realize that the panel drawing contains only information that pertains to that panel and nothing else. Conflicts within that panel are to be documented only on that drawing, but conflicts with other drawings shall be established by the site where you work.

Panel drawings show where devices are mounted and the associated wiring for that panel only. Panels are often manufactured as separate pieces of equipment that are brought to the job site when completely finished. The drawing package that accompanies such a panel must contain, within itself, all necessary information for understanding the function of every device on the panel.

18.8 Plot Plans

We are used to plot plans in the electrical industry. These plans show us where pieces of equipment are located and their relationship to others.

An instrument plot plan is no different. Plot plans show associated instruments that are to be categorized by Tag Number. Tag Numbers on the drawings reflect instruments in reference to their location.

A plot plan is a handy tool for figuring out where a device is to be installed by area only. Elevation details, installation details, loop-wiring methods, etc. are not shown on plot plans. Once the area of a device is known, you must refer to other documentation to receive more details.

There is one other consideration that must be remembered when using plot plans. Some engineering companies only show on plot plans those devices that have both piping and electrical connections. Devices such as a rotometer, which is a type of flow meter that is locally monitored, will only show up on installation details and process diagrams, but not on plot plans.

In today's environment, there are often two sets of plot plans: one showing all devices with electrical connections and one showing devices with associated piping or pneumatic connections. Care must be taken to avoid any conflicts between the two drawings, for often, as we now know, devices have both electrical and pneumatic connections and will be shown on two drawings. If a conflict arises, the site where you work must establish hierarchy because there is no standard for plot plans.

18.9 P&IDs

Process and instrumentation diagrams (drawings), commonly abbreviated to P&IDs, are considered the primary source of reference for

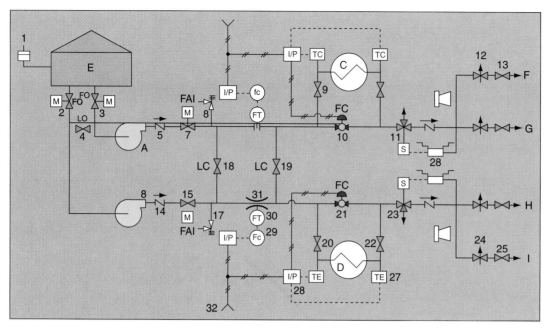

FIGURE 18–6 Example P&ID.

documentation of control systems. P&IDs are usually designed to present functional information about a system or component. Examples are piping layout, flow paths, pumps, valves, instruments, signal modifiers, and controllers, as illustrated in Figure 18–6.

As a rule, P&IDs do not have a drawing scale and present only the relationship or sequence between components. Just because two pieces of equipment are drawn next to each other does not indicate that, in the plant, the equipment is even in the same building; it is just the next part or piece of the system. These drawings only present information on how a system functions, not the actual physical relationships. Because P&IDs provide the most concise format for how a system should function, they are used extensively in the operation, repair, and modification of the plant.

P&IDs contain information about the process, piping, relevant pieces of equipment, control and monitoring instrumentation, process reactions, temperatures, and anything else that could be considered relevant to the process and/or control of the process.

Such an excessive amount of information packed into such a relatively small drawing must employ the use of symbols to maintain order among such limited space. P&IDs often are split into two separate drawings: one that shows the mechanical relationships to the process and another that shows the electrical relationship. When two drawings are used, it is good practice to combine them into one drawing, when possible, to avoid future conflicts.

P&IDs present a pictorial representation of a process. Often, several P&IDs are needed to fully picture an entire process and, from there, it is easy to picture that the number needed to depict an entire industrial plant will be in the hundreds. The volume of P&IDs that must be kept up to date and reviewed requires a consistent approach to representation.

FIGURE 18–7 Miscellaneous symbols.

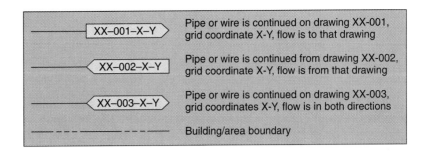

XX–001–X–Y	Pipe or wire is continued on drawing XX-001, grid coordinate X-Y, flow is to that drawing
XX–002–X–Y	Pipe or wire is continued from drawing XX-002, grid coordinate X-Y, flow is from that drawing
XX–003–X–Y	Pipe or wire is continued on drawing XX-003, grid coordinates X-Y, flow is in both directions
	Building/area boundary

The elevations of different pieces of equipment are related. The process line sizes represent the magnitude of flow. In general, the physical machinery, vessels, piping, etc. used are drawn to "look" how they will be in the field. It is relatively easy to pick out a pump or a burner from the P&ID, but the instrumentation may present a slightly different problem.

The symbology becomes much more critical with instrumentation. Instrument devices are shown by their appropriate symbol and the installer must be able to recognize them. Again, devices are categorized by their respective Tag Numbers.

Tag Numbers, along with the type of service, are depicted on the P&ID and, from there, the instrument worker must determine what is needed. Technicians use the P&ID to troubleshoot by observing a device's function with respect to the process. If a process does not respond properly, the P&ID can often provide a hint to the underlying problem.

For today's control systems, the controller that is shown on the P&ID will be shown as a "black box." This is because electronic control methods and/or logic are not to be shown on the drawing to eliminate control confusion. Electronic and, therefore, microprocessor-controlled methods are separate controlling methods that are not to be confused with the process. Often enough, several control schemes may be run, with each using the same instrumentation located on the same pieces of equipment to make different products by only changing the "logic" of the process.

Some P&IDs depict setpoints and transmitter ranges, but these are the exception rather than the norm. When a conflict does occur between the P&ID and any other source of documentation, the P&ID usually has authority. Always follow the proper documentation hierarchy for the location where you work.

In addition to the normal symbols used on P&IDs to represent specific pieces of equipment, miscellaneous symbols are used to guide or provide additional information about the drawing. Figure 18–7 lists and explains four of the more common miscellaneous symbols.

18.10 Isometric Drawings

The isometric projection presents a single view of the component or system. The view is commonly from above and at an angle of 30°. This provides a more realistic, three-dimensional view. As shown in Figure 18–8, this view makes it easier to see how the system looks and how its various portions or parts are related to one another. Isometric projections may or may not be drawn to scale.

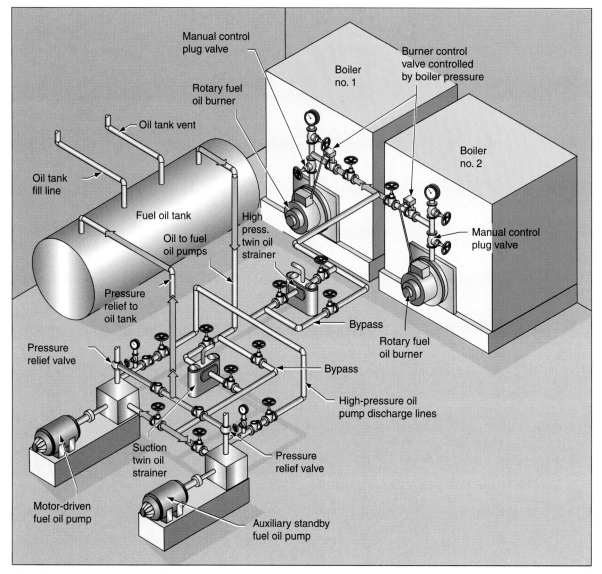

FIGURE 18–8 Isometric drawing.

18.11 Calibration Data Sheets

Calibration (cal) data sheets are used when calibrating an instrument to record the actual values applied and data response of the device calibrated. The calibration data sheet will be the record used to determine if a device is calibrated correctly, the equipment used during calibration, and the person performing the calibration. Calibration data sheets are used to create a permanent device history. The evidence obtained from a cal sheet will often provide maintenance scheduling information. They will also be used to help create the specification sheets, which are a permanent device specifications list.

An example calibration data sheet is shown in Figure 18–9. Data recorded in a "cal sheet" has to be recorded accurately and completely. For example, if a procedure to calibrate a differential pressure device was complete and accurate, the data entered into the input column should be of sufficient "decimal points" to ensure the required accu-

FIGURE 18–9 Calibration data record.

Manufacturer:		
Instrument ID:		Model:
Calibration Range:		
Input:		Output:

Test Equipment	Model	S/N

Accuracy %:

%	Input	Desired	As Found	As Left

Performed By:

Date:

racy of the device. Likewise, the output signal should be listed in mA to sufficient decimal points to ensure accuracy. It is important that the "engineering units" be entered into the cal sheet as well. It is important to note, as covered earlier, that the communicator for a smart device is not listed as calibration equipment used for the calibration procedure.

■ SUMMARY

This chapter discussed the drawings used by technicians to help them install and maintain a control system. Each drawing plays a crucial role in the installation process. This chapter provided the purpose of each drawing that is available. The drawing system discussed in this chapter is not the only system. Readers should also be aware of other types of drawings available for their use.

■ REVIEW QUESTIONS

1. A generic drawing can be divided into five major parts. List them below.
2. What do "clouds," or scoped, areas indicate?
3. Where should drawing symbols and notes always be referenced?
4. When are instrument index sheets created? For what purpose?
5. Which documentation sheet is available as a permanent record of the device after installation and "turn-over"?
6. What documentation displays the device, wiring path, termination points, and power supply?
7. What documentation should be completed to record the actual calibration process?

chapter 19

Fundamentals of Safety in the Process Environment

■ OUTLINE

■ OVERVIEW

This chapter presents material provided by the OSHA Safety and Heath Construction Standards Excerpts. The material presented in this chapter is intended to create a safer, more effective workforce. This chapter is not intended to cover all of the safety concerns that anyone working in the process environment may come into contact with, but it does provide a general safety reminder for all projects, in all facilities.

■ OBJECTIVES

After completing this chapter, you should be able to:

■ Recall and implement the correct procedures for lockout/tagout requirements.
■ List the safety concerns for excavation work requirements.
■ Provide a list of requirements to prevent falls on the job.
■ Explain the safety concerns associated with working in confined spaces.
■ Demonstrate the proper concerns for working from ladders.

■ SAFETY IN THE PROCESS ENVIRONMENT

19.1 Lockout/Tagout Procedures

Osha Safety & Health Construction Standards Excerpts

Electrical Safety

Safe Work Practices—Subpart K

Lockout/Tagout

■ Controls that are to be deactivated during the course of work on energized or deenergized equipment or circuits shall be tagged.
■ Equipment or circuits that are deenergized shall be rendered inoperative and shall have tags attached at all points where such equipment or circuits can be energized.
■ Tags shall be placed to identify plainly the equipment or circuits being worked on.
■ While any employee is exposed to contact with parts of fixed electric equipment or circuits which have been deenergized, the circuits energizing the parts shall be locked out or tagged or both in accordance with the OSHA requirements listed below. The requirements shall be followed in the order in which they are presented.
  1. The employer shall maintain a written copy of their lockout/tagout procedures and shall make it available for inspection by employees and by the Assistant Secretary of Labor and his or her authorized representatives;
  2. Safe procedures for deenergizing circuits and equipment shall be determined before circuits or equipment are deenergized.
  3. The circuits and equipment to be worked on shall be disconnected from all electric energy sources. Control circuit devices such as push buttons, selector switches, and interlocks may not be used as the sole means for deenergizing circuits or equipment. Interlocks for electric equipment may not be used as a substitute for lockout and tagging procedures.

4. Stored electric energy which might endanger personnel shall be released. Capacitors shall be discharged and high-capacitance elements shall be short-circuited and grounded, if the stored electric energy might endanger personnel.

5. Stored nonelectrical energy in devices that could reenergize electric circuit parts shall be blocked or relieved to the extent that the circuit parts could not be accidentally energized by the device.

6. A lock and a tag shall be placed on each disconnecting means used to deenergize circuits and equipment on which work is to be performed. The lock shall be attached so as to prevent persons from operating the disconnecting means unless they resort to undue force or the use of tools.

7. Each tag shall contain a statement prohibiting unauthorized operation of the disconnecting means and removal of the tag.

8. If a lock cannot be applied, or if the employer can demonstrate that tagging procedures will provide a level of safety equivalent to that obtained by the use of a lock, a tag may be used without a lock.

9. Where a tag is permitted to be used without a lock, at least one additional safety measure that provides a level of safety equivalent to that obtained by use of a lock shall be employed. Examples of additional safety measures include the removal of an isolating circuit element, blocking of a controlling switch, or opening of an extra disconnecting device.

10. A lock may be placed without a tag only under the following conditions:

    a. Only one circuit or piece of equipment is deenergized, and

    b. The lockout period does not extend beyond the work shift, and

    c. Employees exposed to the hazards associated with reenergizing the circuit or equipment are familiar with this procedure.

11. The requirements listed above shall be met before any circuits or equipment can be considered and worked as deenergized.

12. A qualified person shall operate the equipment-operating controls or otherwise verify that the equipment cannot be restarted.

13. A qualified person shall use test equipment to test the circuit elements and electrical parts of equipment to which employees will be exposed and shall verify that the circuit elements and equipment parts are deenergized. The test shall also determine if any energized condition exists as a result of inadvertently induced voltage or unrelated voltage backfeed even though specific parts of the circuit have been deenergized and presumed to be safe. If the circuit to be tested is over 600 volts, nominal, the test equipment shall be checked for proper operation immediately after this test.

The following requirements shall be met, in the order given, before circuits or equipment are reenergized, even temporarily.

1. A qualified person shall conduct tests and visual inspections, as necessary, to verify that all tools, electrical jumpers, shorts, grounds, and other such devices have been removed, so that the circuits and equipment can be safely energized.
2. Employees exposed to the hazards associated with reenergizing the circuit or equipment shall be warned to stay clear of circuits and equipment.
3. Each lock and tag shall be removed by the employee who applied it or under his or her direct supervision.
4. There shall be a visual determination that all employees are clear of the circuits and equipment.

19.2 Excavations

OSHA Safety & Health Construction Standards Excerpts

Excavations

- The estimated location of utility installations, such as sewer, telephone, fuel, electric, water lines, or any other underground installations that reasonably may be expected to be encountered during excavation work shall be determined prior to opening an excavation.
- A stairway, ladder, ramp, or other safe means of egress shall be located in trench excavations that are 4 feet or more in depth so as to require no more than 25 feet of lateral travel for employees.
- No employee shall be permitted underneath leads handled by lifting or digging equipment. Employees shall be required to stand away from any vehicle being loaded or unloaded to avoid being struck by any spillage or falling materials.
- Where oxygen deficiency (atmospheres containing less than 19.5 percent oxygen) or hazardous atmosphere exists or could reasonably be expected to exist, such as in excavations in landfill areas or excavations in areas where hazardous substances are stored nearby, the atmospheres in the excavation shall be tested before employees enter excavations greater than 4 feet in depth.
- Emergency rescue equipment, such as breathing apparatus, a safety harness and line, or a basket stretcher, shall be readily available where hazardous atmospheric conditions exist or may reasonably be expected to develop during work in an excavation.
- Employees shall not work in excavations in which there is accumulated water, unless adequate precautions have been taken to protect employees against the hazards posed by water accumulation. The precautions necessary to protect employees adequately vary with each situation, but could include special support or shield systems to protect from cave-ins, water removal to control the level of accumulating water, or use of a safety harness and lifeline.
- Employees shall be protected from excavated or other materials or equipment that could pose a hazard by falling or rolling into excavations. Protection shall be provided by placing and keeping such

materials or equipment at least 2 feet from the edge of excavations, or by the use of retaining devices that are sufficient to prevent materials or equipment from falling or rolling into excavations, or by a combination of both if necessary.

19.3 Fall Protection

OSHA Safety & Health Construction Standards Excerpts

Fall Protection

- Each employee on a walking/working surface (horizontal and vertical surface) with an unprotected side or edge that is 6 feet or more above a lower level shall be protected from falling by the use of guardrail systems, safety net systems, or personal fall arrest systems.

- Each employee who is constructing a leading edge 6 feet or more above lower levels shall be protected from falling by guardrail systems, safety net systems, or personal fall arrest systems.

- Each employee on a walking/working surface 6 feet or more above a lower level where leading edges are under construction, but who is not engaged in the leading edge work, shall be protected from falling by a guardrail system, safety net system, or personal fall arrest system. If a guardrail system is chosen to provide the fall protection, and a controlled access zone has already been established for leading edge work, the control line may be used in lieu of a guardrail along the edge that parallels the leading edge.

- Each employee in a hoist area shall be protected from falling 6 feet or more to lower levels by guardrail systems or personal fall arrest systems. If guardrail systems (or chain, gate, or guardrail), or portions thereof, are removed to facilitate the hoisting operation (e.g., during landing of materials), and an employee must lean through the access opening or out over the edge of the access opening (to receive or guide equipment and materials, for example), that employee shall be protected from fall hazards by a personal fall arrest system.

- Each employee on walking/working surfaces shall be protected from falling through holes (including skylights) more than 6 feet above lower levels, by personal fall arrest holes. All covers shall be color coded or they shall be marked with the word "HOLE" or "COVER" to provide warning of the hazard.

- Each employee on a walking/working surface shall be protected from tripping in or stepping into or through holes (including skylights) by covers.

- Each employee on a walking/working surface shall be protected from objects falling through holes (including skylights) by covers.

- Each employee on ramps, runways, and other walkways shall be protected from falling 6 feet or more to lower levels by guardrail systems.

- Each employee at the edge of an excavation 6 feet or more in depth shall be protected from falling by guardrail systems, fences, or barricades when the excavations are not readily seen because of plant growth or other visual barriers.

- Each employee at the edge of a well, pit, shaft, and similar excavation 6 feet or more in depth shall be protected from falling by guardrail systems, fences, barricades, or covers.

- Each employee less than 6 feet above dangerous equipment shall be protected from falling into or onto the dangerous equipment by guardrail systems or by equipment guards. Each employee 6 feet or more above dangerous equipment shall be protected from fall hazards by guardrail systems, personal fall arrest systems, or safety net systems.

- Each employee reaching more than 10 inches below the level of the walking/working surface on which they are working shall be protected from falling by a guardrail system, safety net system, or personal fall arrest system.

- Each employee working on, at, above, or near wall openings (including those with chutes attached) where the outside bottom edge of the wall opening is 6 feet or more above lower levels and the inside bottom edge of the wall opening is less than 39 inches above the walking/working surface shall be protected from falling by the use of a guardrail system, a safety net system, or a personal fall arrest system.

- When an employee is exposed to falling objects, the employer shall have each employee wear a hard hat and shall implement one of the following measures:

  1. Erect toeboards, screens, or guardrail systems; or,

  2. Erect a canopy structure and keep potential fall objects far enough from the edge of the higher level so that those objects would not go over the edge if they were accidentally displaced; or,

  3. Barricade the area to which objects could fall, prohibit employees from entering the barricaded area, and keep objects that may fall far enough away from the edge of a higher level so that those objects would not go over the edge.

19.4 Confined Spaces

OSHA Safety & Health General Industry Standards Excerpts

Confined Spaces

The employer shall ensure that each attendant:

1. Knows the hazards that may be faced during entry, including information on the mode, signs or symptoms, and consequences of the exposure;

2. Is aware of possible behavioral effects of hazard exposure in authorized entrants;

3. Continuously maintains an accurate count of authorized entrants in the permit space;

4. Remains outside the permit space during entry operations until relieved by another attendant;

5. Communicates with authorized entrants as necessary to monitor entrant status and to alert entrants of the need to evacuate the space;

6. Monitors activities inside and outside the space to determine if it is safe for entrants to remain in the space and orders the authorized entrants to evacuate the permit space immediately under hazardous conditions;

7. Summons rescue and other emergency services as soon as the attendant determines that authorized entrants may need assistance to escape from permit space hazards;

8. Keeps unauthorized persons from approaching or entering a permit space while entry is underway;

9. Performs *non-entry* rescues as specified by the employer's rescue procedure; and

10. Performs no duties that might interfere with the attendant's primary duty to monitor and protect the authorized entrants.

   ■ Each authorized entrant shall use a chest or full body harness, with a retrieval line attached at the center of the entrant's back near shoulder level, or above the entrant's head. Wristlets may be used in lieu of the chest or full body harness if the employer can demonstrate that the use of a chest or full body harness is infeasible or creates a greater hazard and that the use of wristlets is the safest and most effective alternative.

   ■ The other end of the retrieval line shall be attached to a mechanical device or fixed point outside the permit space in such a manner that rescue can begin as soon as the rescuer becomes aware that rescue is necessary. A mechanical device shall be available to retrieve personnel from vertical-type permit spaces more than 5 feet deep.

19.5 Ladder Safety

Safety Extra: Ladder Safety

Stay on the Ladder

■ Each year, hundreds of workers are killed by falls while working from ladders. Twice as many falls occur when stepping down compared to going up ladders. The main cause of falls from straight and extension ladders is sliding of the ladder base. For self-supported ladders or stepladders, the main cause is tipping sideways.

■ Make sure the ladder is about 1 foot away from the vertical support for every 4 feet of ladder height between the foot and the top support (Figure 19–1).

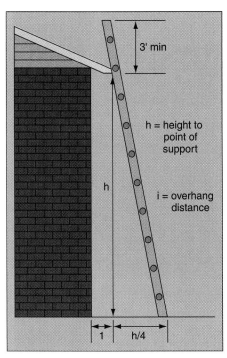

FIGURE 19–1 Dimensions for ladder safety.

Table 19-1   Load Capacity Ratings

Load Capacity Ratings	
375 lb Type IAA duty rating	Special duty/professional use
300 lb Type IA duty rating	Extra heavy duty/professional use
250 lb Type I duty rating	Heavy duty/industrial use
225 lb Type II duty rating	Medium duty/commercial use
200 lb Type III duty rating	Light duty/household use

Five Rules of Ladder Safety

1. Select the right ladder for the job:
   - There are many types of ladders intended for specific purposes—be careful when placing a ladder around electrical wires. *Do not use aluminum.*
2. Inspect the ladder before you use it:
   - Each time you use a ladder, inspect it for loose or damaged rungs, rails, or braces. Repair defects before use.
3. Set up the ladder with care:
   - An improperly set up ladder adds to the risk of an accident. Stepladders should be fully opened and straight ladders set up based on the height at support. A minimum of 3 feet of ladder should extend above the edge of support.
4. Climb and descend ladders cautiously:
   - Always face the ladder and climb with both hands. Carry tools in a belt or raise and lower them with a handline.
5. Use common sense when working on a ladder:
   - Hold on with one hand and never reach too far to either side. Never climb higher than the second step from the top of a stepladder or the third from the top of a straight ladder.

19.6 Summarizing Safety-Related Work Practices

This section provides a list of references that address safety concerns that may be present in the installation of control systems or devices. Each reference provided here is not all-encompassing, but rather is a general reference that may be researched further. The following references of the National Electrical Code (NEC), National Fire Protection Association (NFPA), Occupational, Safety and Heath Administration (OSHA) are a field and course of study alone. Use the following information as a source of reference only to assist in understanding the safety requirements that are in place today for anyone who is present during construction or maintenance of the control system.

The following information is not to replace or take from "site-specific" safety training. Each facility often has additional training that is required to be completed before work may be performed.

References

NEC:

Section 110.16 of the 2005 National Electrical Code has a fine print note stating:

> NFPA 70E-2004, Standard for Electrical Safety in the Workplace, provides assistance in determining severity of potential exposure, planning safe work practices, and selecting personal protective equipment.

This reference of an electrical safety standard is to ensure that safe work practices are followed and observed by all who participate in the installation of a control system or single devices.

OSHA:

Safety-related work practice requirements, such as:
For Construction . . .

> No employer shall permit an employee to work in such proximity to any part of an electric power circuit that the employee could contact the electric power circuit in the course of work, unless the employee is protected against electric shock by deenergizing the circuit and grounding it or by guarding it effectively by insulation or other means.

For General Industry . . .

> Safety-related work practices shall be employed to prevent electric shock or other injuries resulting from either direct or indirect electrical contacts, when work is performed near or on equipment or circuits which are or may be energized. The specific safety-related work practices shall be consistent with the nature and extent of the associated electrical hazards.

■ SUMMARY

This chapter provided general safety requirements for anyone who is working in the process controls area. There are many more areas of concern, and facilities often require additional safety training. This chapter presented areas of concern more than specific applications of safety requirements. The required and relevant safety precautions are always provided by the facility where the work is being performed.

■ REVIEW QUESTIONS

1. As a general rule, when should equipment energizing circuits and/or equipment be locked out?
2. Can control devices such as push buttons and selector switches be used for deenergizing circuits or equipment?
3. Who is authorized to remove each lock and tag applied?
4. What is the permitted depth of an excavation before an exit means is required for personnel working in the excavation?

Appendix A— Instrumentation and Controls Symbology

The symbols listed in this appendix are supplied as supplemental material to aid and assist in interpreting instrumentation drawings. The following are not intended as a set of complete references for symbols, but they do provide an enhanced reference for the reader. Relevant symbols are identified by subject/topic headings.

Table A–1   Control Valve Symbols

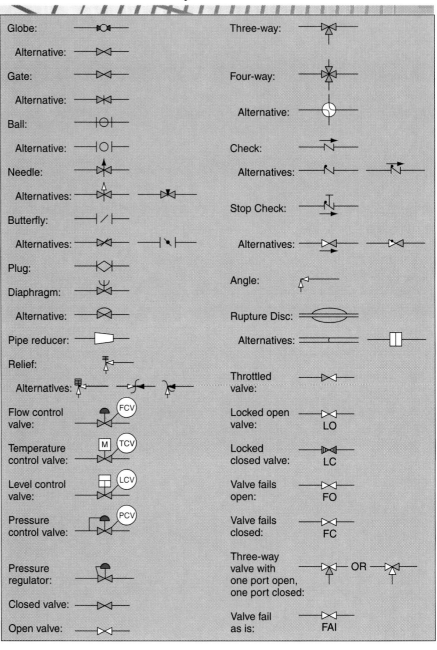

Table A–2 Control Valve Actuator Symbols

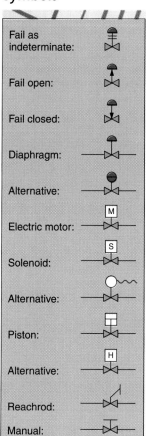

Fail as indeterminate:	
Fail open:	
Fail closed:	
Diaphragm:	
Alternative:	
Electric motor:	
Solenoid:	
Alternative:	
Piston:	
Alternative:	
Reachrod:	
Manual:	

Table A–3 Contacts and Symbols

Limit switch, direct actuated, spring returned normally open	
Normally open-held closed	
Normally closed	
Normally closed-held open	
Open switch with time-delay closing (TDC) feature	TDC OR
Closed switch with time-delay opening (TDO) feature	TDO OR
Open switch with time-delay opening (TDO) feature	TDO OR
Closed switch with time-delay closing (TDC) feature	TDC OR
Flow-actuated switch closing on increase in flow	
Opening on increase in flow	
Liquid level actuated switch closing on rising level	
Opening on rising level	
Pressure or vacuum-actuated switch closing on rising pressure	
Opening on rising pressure	
Temperature-actuated switch closing on rising temperature	IEC OR

(continued)

Table A–3 Continued

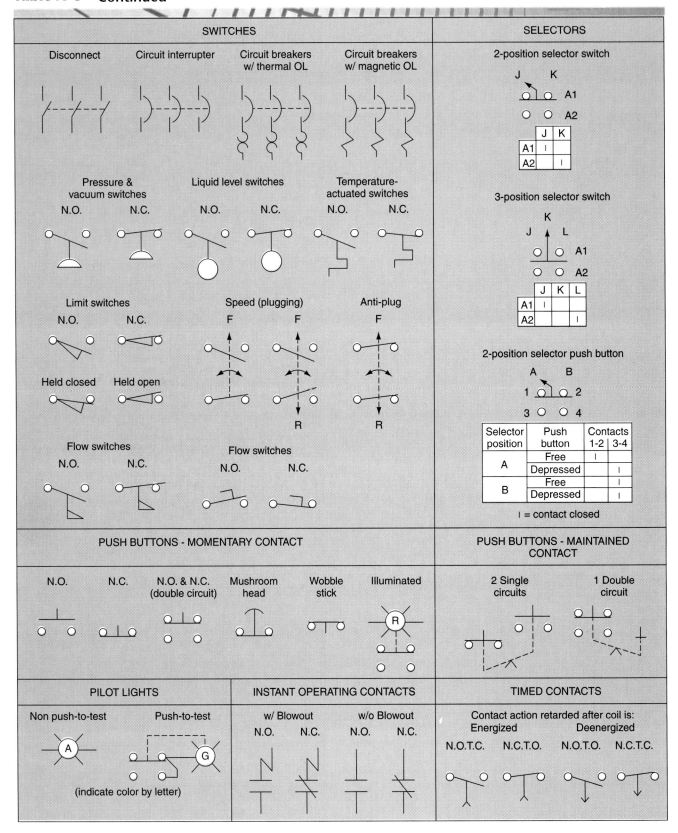

Table A–3   Continued

(continued)

Table A–3   Continued

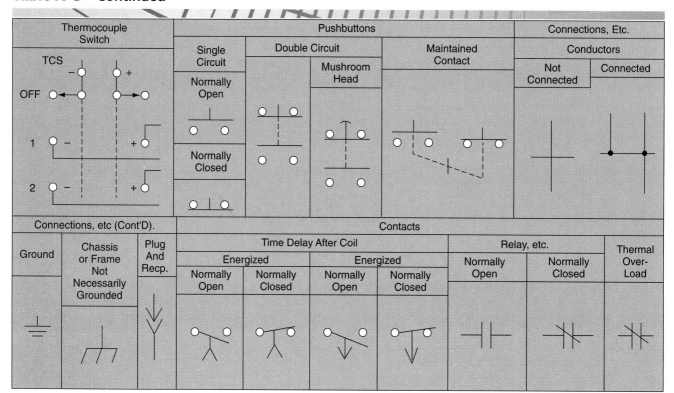

Table A–4   Pipe Fitting Valve Symbology

FITTINGS & VALVES		
Bushing		Street elbow
Cap		Joint, connecting pipe
Cross, reducing		Expansion joint
Cross, straight size		Lateral
Cross-over		Pipe plug
45° Elbow		Concentric reducer
90° Elbow		Eccentric reducer
Elbow turned down		Sleeve
Elbow turned up		Tee, straight size
Base elbow		Tee, outlet up
Double-branch elbow		Tee, outlet down
Long-radius elbow		Tee, double sweep
Reducing elbow		Tee, reducing
Side outlet elbow (outlet down)		Tee, single sweep
Side outlet elbow (outlet up)		Tee, side outlet (outlet down)
		Tee, side outlet (outlet up)

NOTE: Pipe fittings and valves may be:				
Flanged	Screwed	Bell & spigot	Welded	Soldered

(continued)

Table A–4 Continued

FITTINGS & VALVES		MISC. VALVES	
―‖―	Union	→	Check valve
ANGLE VALVES			Cock valve
	Check		Diaphragm valve
	Gate (elevation)		Float valve
	Gate (plan)		Gate valve
	Globe (elevation)	M	Motor operated
	Globe (plan)		Globe valve
	Hose angle	M	Globe valve, (motor operated)
			Hose valve, gate
			Hose valve, globe
			Lock shield valve
			Quick opening valve
			Safety valve

Table A–5   Resistor Color-Code Chart

1ST AND 2ND COLOR BANDS		DIGIT IT REPRESENTS	MULTIPLIER
	Black	0	× 1
	Brown	1	× 10
	Red	2	× 100
	Orange	3	× 1,000 or 1K
	Yellow	4	× 10,000 or 10K
	Green	5	× 100,000 or 100K
	Blue	6	× 1,000,000 or 1M
	Violet	7	Silver is divided by 100
	Gray	8	Gold is divided by 10
	White	• 9	Tolerances: • Gold = 5% • Silver = 10% • None = 20%

Table A–6   100-Ohm RTD Chart

Resistance vs. temperature for a 100-ohm platinum RTD
ALPHA = 0.00385
DIN 43760
BS 1904.

°F	°C	Ohms
1,000	537.78	293.49
900	482.22	275.04
800	426.67	256.23
700	371.11	237.07
600	315.56	217.55
500	260.0	197.69
400	204.44	177.47
300	148.89	156.90
200	93.33	135.97
150	65.56	125.37
100	37.78	114.68
50	10.0	103.90
32	0.0	100.00
0	−17.78	93.03
−50	−45.56	82.06
−100	−73.33	70.98

Table A–7   Electronic Symbols

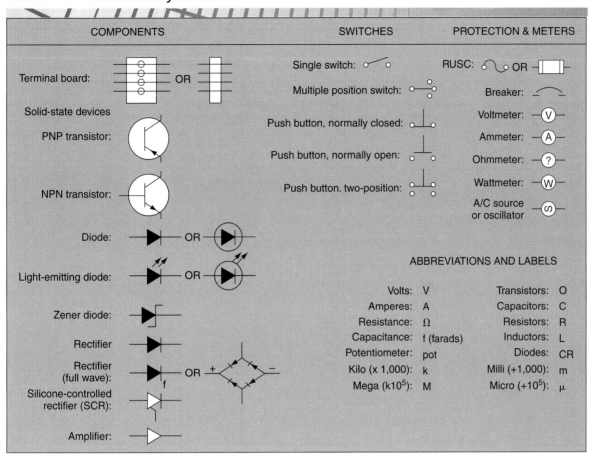

COMPONENTS	SWITCHES	PROTECTION & METERS

COMPONENTS

Terminal board: OR

Solid-state devices

PNP transistor:

NPN transistor:

Diode: OR

Light-emitting diode: OR

Zener diode:

Rectifier:

Rectifier (full wave):   OR

Silicone-controlled rectifier (SCR):

Amplifier:

SWITCHES

Single switch:

Multiple position switch:

Push button, normally closed:

Push button, normally open:

Push button. two-position:

PROTECTION & METERS

RUSC: OR

Breaker:

Voltmeter: V

Ammeter: A

Ohmmeter: ?

Wattmeter: W

A/C source or oscillator:

ABBREVIATIONS AND LABELS

Volts:	V	Transistors:	O
Amperes:	A	Capacitors:	C
Resistance:	Ω	Resistors:	R
Capacitance:	f (farads)	Inductors:	L
Potentiometer:	pot	Diodes:	CR
Kilo (x 1,000):	k	Milli (+1,000):	m
Mega (k10^5):	M	Micro (+10^5):	μ

Table A–8 PLC and Motor Control Symbols

PROTECTION DEVICES	OUTPUT DEVICES
Fuse	Light indicator
Fuse	Motor
Overload	DC motor armature
Overload thermal	3-phase motor
Overload magnetic	Heater
OL'S Ladder logic overload	LED indicator
	Solenoid

Table A–9 PLC & Motor Control Symbology

SWITCHES & CONTACTS	TRANSFORMERS
2-Position single-throw	Coil
2-Position double-throw	Air core transformer
Rotary, multiposition	Iron core transformer
2-Pole single-throw (dashed lines indicate contacts mechanically not electrically connected)	Variable transformer
2-Pole double-throw	Auto transformer
Manual toggle switch	Single-phase power 1 line drawing
(NO) Push button switch	3-Phase delta transformer
(NO) Manual foot switch	OR Current transformer
(NC) Limit switch	Potential transformer
(NO) Limit switch	
(NC) Held open limit switch	**COILS AND RELAYS**
(NO) Held closed limit switch	
Paddle flow switch	(M) Motor starter coil
Float switch	(F) Forward or fast starter coil
Pressure or vacuum switch	(R) Reverse starter coil
Time-delay switch	(S) Slow starter coil
Temperature switch	(CR) Control relay
Contacts (NO)	(TDR) Time delay relay NOTE: Various letter abbreviations may appear in the circles above.
Contacts (NC)	

Table A–10 General Conversion Factors

To Obtain	Multiply	By
Atmospheres	In HG@32°F	0.033421
BTU	Watt-hours	3.412
BTU	KWh	.3412
Centimeters	inches	2.54°
Cm of HG @ 0 deg C	Atmospheres	76.0
Cm of HG @ 0 deg C	Grams/sq. cm	0.07356
Cm of HG @ 0 deg C	Lb/sq in.	5.1715
Cm of HG @ 0 deg C	Lb/sq ft	0.035913
Cm/(sec)(sec)	Gravity	980.665
Centipoises	Centistokes	Density
Centistokes	Centipoises	1/density
Cu cm	Cu ft	28,317.0
Cu cm	Cu in.	16–387
Cu cm	Gal (USA, liq.)	3785.43
Cu cm	Liters	1000.03
Cu cm	Quarts (USA, liq.)	946.358
Cu cm/sec	Cu ft/min	472.0
Cu ft	Cu meters	35.314
Cu ft	Gal (USA, liq.)	0.13368
Cu ft	Liters	0.03532
Cu ft/min	Cu meters/sec	2118.9
Cu ft/min	Gal (USA, liq.)/sec	8.0192
Cu ft/sec	Gal (USA, liq.)/min	0.0022280
Cu ft/sec	Liters/min	0.0005886
Cu in.	Cu centimeters	0.061023
Cu in.	Gal (USA, liq.)	231.0
Cu in.	Liters	61.03
Cu meters	Gal (USA, liq.)	0.0037854
Cu meters	Liters	0.001000028
Cu meters/hr	Gal/min	0.22712
Cu meters/kg	Cu ft/lb	0.062428
Cu meters/min	Cu ft/min	0.02832
Cu meters/sec	Gal/min	0.000063088
Feet	Meters	3.281
Ft/min	Cm/sec	1.9685
Ft/sec	Meters/sec	3.2808
Ft/(sec)(sec)	Gravity (sea level)	32.174
Ft/(sec)(sec)	Meters/(sec)(sec)	3.2808
Gal (Imperial, liq.)	Gal (USA, liq.)	0.83268
Gal (USA, liq.)	Barrels (Petroleum, USA)	42.0
Gal (USA, liq.)	Cu ft	7.4805
Gal (USA, liq.)	Cu meters	264.173
Gal (USA, liq.)	Cu yards	202.2
Gal (USA, liq.)	Gal (Imperial, liq.)	1.2010
Gal (USA, liq.)	Liters	0.2642
Gal (USA, liq.)/min	Cu ft/sec	448.83
Gal (USA, liq.)/min	Cu meters/hr	4.4029
Gal (USA, liq.)/sec	Liters/min	0.0044028
Grams	Pounds (a voir.)	453.5924

Table A–11   Temperature Conversion Table

Locate known temperature in °C/°F column. Read converted temperature in °C or °F column.

°C	°C/°F	°F	°C	°C/°F	°F	°C	°C/°F	°F
−45.4	−50	−58	15.5	60	140	76.5	170	338
−42.7	−45	−49	18.3	65	149	79.3	175	347
−40	−40	−40	21.1	70	158	82.1	180	356
−37.2	−35	−31	23.9	75	167	85	185	365
−34.4	−30	−22	26.6	80	176	87.6	190	374
−32.2	−25	−13	29.4	85	185	9.4	195	383
−29.4	−20	−4	32.2	90	194	93.2	200	392
−26.6	−15	5	35	95	203	96	205	401
−23.8	−10	14	37.8	100	212	98.8	210	410
−20.5	−5	23	40.5	105	221	101.6	215	419
−17.8	0	31	43.4	110	230	104.4	220	428
−15	5	41	46.1	115	239	107.2	225	437
−12.2	10	50	48.9	120	248	110	230	446
−9.4	15	59	51.6	125	257	112.8	235	455
−6.7	20	68	54.4	130	266	115.6	240	464
−3.9	25	77	57.1	135	275	118.2	245	473
−1.1	30	86	60	140	284	120.9	250	482
1.7	35	95	62.7	145	293	123.7	255	491
4.4	40	104	65.5	150	302	126.5	260	500
7.2	45	113	68.3	155	311	129.3	265	509
10	50	122	71	160	320	132.2	270	518
12.8	55	131	73.8	165	329	135	275	527

$°F = (9/5 \times °C) + 32$

$°C = 5/9 \, (°F − 32)$

Appendix B—NJATC Instrumentation and Process Control Training System

Please refer to the CD enclosed with this book for illustrations.

Absolute pressure   Gauge pressure plus atmospheric pressure.

Absolute zero   Temperature at which thermal energy is at a minimum. Defined as 0° K.

Accessible   Describing a device or function that can be used or seen by an operator for the purpose of performing control actions.

Accuracy   Conformity to an indicated, standard, or true value, usually expressed as a percentage (of span, reading, or upper-range value) deviation from the indicated, standard, or true value.

Actuator   A part of the final control element that converts a signal into a forced action of the final control element.

Alarm   A device or function that signals the existence of an abnormal condition by means of audible or visible discrete changes, or both, intended to attract attention.

Algorithm   A detailed set of instructions that is to be executed by the CPU.

Alphanumeric   A character set that contains both letters and digits.

Ambient compensation   (1) The design of an instrument such that changes in ambient temperature do not affect readings of the instrument. (2) Compensation for ambient conditions when mounting an instrument.

Ambient conditions   Conditions around the device examined (pressure, temperature, etc.).

Ammeter   An instrument used to measure current.

Ampere   A unit used to define the rate of current flow (1 amp = 1 coulomb per second).

Amplification   The dimensionless ratio of output/input in a device intended by design to have this ratio greater than unity.

Amplifier   A device whose output by design is an enlarged reproduction of the input signal and is energized from a source other than the input.

Amplitude   A measurement of the distance from the highest to the lowest excursion of motion; for example, the peak-to-peak swing of an electrical waveform.

Amplitude ratio   The ratio of the magnitude of a steady-state sinusoidal output with respect to the input.

Analog   A continuous operation signal.

Analog computer   A computer operating on continuous variables.

Analog output   A voltage or current signal that is a continuous function of the measured parameter.

Analog simulator   An electronic, pneumatic, or mechanical device that solves problems by simulation of the physical system under study using electrical or physical variables to represent the process variables.

Analog-to-digital converter (A/D)   A device or circuit that outputs a binary number corresponding to an analog signal level at the input.

Analyzer   An automatically operating, analytical measuring device that monitors a process for one or more chemical compositions and/or physical properties.

Anemometer   An instrument for measuring and/or indicating the velocity of air flow.

ANSI   American National Standards Institute.

ASCII   American Standard Code for Information Interchange. A seven- or eight-bit code used to represent alphanumeric characters.

Assignable   A term applied to a feature permitting the channeling (or directing) of a signal from one device to another without the need for switching, patching, or wiring changes.

Attenuation   A decrease in signal magnitude; the reciprocal of *gain*.

Auto-manual station   Synonym for *control station*.

Auto-zero   An automatic, internal correction for offsets and/or drift at zero voltage input.

Automatic control system   An operable arrangement of one or more automatic controllers along with their associated equipment connected in closed loops with one or more processes.

Automatic controller   A device, or combination of devices, that measures the value of the variable, quantity, or condition and operates to correct or limit deviation of this measured value from a selected reference.

Automatic reset   The integral function on a PID controller that adjusts the proportional bandwidth with respect to the setpoint to compensate for droop in the circuit (i.e., adjusts a controlled process to a setpoint after the system stabilizes).

Automation   The act or method of making a processing or manufacturing system without the necessary operator intervention.

Background noise   The total noise floor from all sources of interference in a measurement system, independent of the presence of a data signal.

Backup   A system, device, file, or facility that can be used as an alternative in case of loss of data.

Balloon   Synonym for *bubble*.

Bandwidth   A symmetrical region around the set-point in which proportional control occurs.

Baud   A unit of data transmission speed equal to the number of bits per second (1,200 baud = 1,200 its/sec).

Behind the panel   A term applied to a location that is within an area that contains (1) the instrument panel, (2) its associated rack-mounted hardware, or (3) is enclosed within the panel.

Bellows   A pressure sensor that converts pressure into a linear displacement.

Beta ratio   The ratio of the diameter of a pipeline constriction to the unconstricted pipe diameter.

Binary   A term applied to a signal or device that has only two discrete positions or states (on/off).

Binary-coded decimal (BCD)   The representation of a decimal number (0–9 base 10) by means of a four-bit binary code.

Bit   Acronym for *binary digit.* The smallest unit of computer information, it is a binary 0 or 1.

Block   A set of things such as words, characters, digits, or parameters handled as a unit.

Board   A synonym for *panel.*

Bode diagram   A plot of log-gain and phase angle value on a log-frequency base for an element, loop, or output transfer function. It also comprises similar functional plots of involved variables.

Boiling point   The temperature at which a substance in the liquid phase transforms to the gaseous phase.

Bourdon tube   A pressure sensor that converts pressure to a displacement; a coiled, flattened tube that is straightened when pressure is applied.

Breakpoint   The point of intersection of two confluent, straight-line segments of a plotted curve.

BTU   British Thermal Unit. The quantity of thermal energy required to raise 1 pound of water 1°F, at its maximum density (1 BTU = 0.293 watt-hours = 252 calories).

Bubble   The circular symbol used to denote and identify the purpose of an instrument or function. It may contain a Tag Number.

Buffer   A storage for data that is used to compensate for a speed difference when transferring data from one device to another.

Bulk storage   An auxiliary memory device with storage capacity many orders of magnitude greater than working memory, for example, disk files, drums, magnetic tape drives.

Burst pressure   The maximum pressure applied to a transducer-sensing element or case without causing leakage.

Bus   One or more conductors that transfer signals or power.

Byte   The representation of a character in binary; eight bits.

Calibration   The process of adjusting an instrument or compiling a deviation chart so that its reading can be correlated to the actual value being measured.

Capacitance   The property that may be expressed as the time integral of flow rate (heat, electric current, etc.) to or from storage divided by the associated potential change.

Capacity   Measure of capability to store liquid volume, mass, heat, information, or any additional form of energy or matter.

Cascade control system   A control system in which the output of one controller is the setpoint of another.

Cavitation   The boiling of a liquid caused by a decrease in pressure rather than by an increase in temperature.

Celsius (centigrade)   A temperature scale defined by 0° at the freezing point and 100° at the boiling point of water at sea level.

Centripetal force   A force exerted on an object moving in a circular path that is exerted inward toward the center of rotation.

CFM   The volumetric flow rate of a liquid or gas in cubic feet per minute.

Character   A letter, digit, or other symbol that is used as the representation of other data. A connected sequence of characters is called a *character string.*

Chatter   The rapid cycling on and off of a relay in a control process.

Chip   An integrated circuit.

Closed loop (feedback loop)   Several automatic control units and the process connected so as to provide a signal path that includes a forward path, a feedback path, and a summing point. The controlled variable is consistently measured, and if it deviates from that which has been prescribed, corrective action is applied to the final element in such direction as to return the controlled variable to the desired value.

Color code   The ANSI-established color code for thermocouple wires in the Negative lead is always red. Color code for base metal thermocouples is yellow for type K, black for type J, purple for type E, and blue for type T.

Communication   Transmission and reception of data among data processing equipment and related peripherals.

Compensation   An addition of specific materials or devices to counteract a known error.

Compiler   A program that translates a higher-level language like "BASIC" or "FORTRAN" into an assembly or machine language that the CPU can execute.

Computer   A device that performs mathematical calculations. It may range from a simple device (such as a slide rule) to a very complicated one (such as a digital computer). In process control, the computer is either an analog mechanism set up to perform a continuous calculation on one or more input signals and to provide an output as a function of time without relying on external assistance (human prompting), or a digital device used in direct control (DDC).

Computing device A device or function that performs one or more calculations or logic operations, or both, and transmits one or more resultant output signals. Sometimes called a *computing relay.*

Conductance The measure of the ability of a solution to carry an electrical current.

Configurable A term applied to a device or system whose functional characteristics can be selected or rearranged through programming or other methods.

Control accuracy The degree of correspondence between the controlled variable and the desired value after stability has been achieved.

Control loop Starts at the process in the form of a measurement or variable, is monitored, and returns to the process in the form of a manipulated variable or "valve position" being controlled by some means.

Control mode The output form or type of control action used by a controller to perform control functions (i.e., on/off, time proportioning, PID).

Control point The value at which the controlled system or process settles out or stabilizes. It may or may not agree with the setpoint applied to the controller.

Control station A manual loading station that also provides switching between manual and automatic control modes of a control loop.

Control system A system in which deliberate guidance or manipulation is used to achieve a prescribed value of variable.

Control valve A device, other than a common, hand-actuated ON-OFF valve or self-actuated check valve, that directly manipulates the flow of one or more fluid process streams.

Controlled system The part of a system under control; the process.

Controller A device having an output that varies to regulate a controlled variable in a specified manner.

Controlling means The elements in a control system that contribute to the required corrective action.

Converter A device that receives information in one form of an instrument signal and transmits an output signal in another form.

Coriolis force A result of centripetal force on a mass moving with a velocity radially outward in a rotating plane.

CPS Cycles per second; the rate or the number of periodic events in one second.

CPU Central processing unit; the portion of a computer that decodes the instructions, performs the actual computations, and keeps order in the execution of programs.

Cross-sectional area With reference to circular pipe used for flow routing, the cross-sectional area is found by the equation: $A = \pi r^2$

Cycling A periodic change in the factor under control, usually resulting in equal excursions above and below the control point.

Damping Progressive reduction in the amplitude of cycling of a system. *Critically damped* describes a system that is damped just enough to prevent overshoot following an abrupt change of input variables.

Data A general term to denote any information that can be processed.

Data base A large amount of data stored in a well-organized manner.

Dead band The change through which the input to an instrument can be varied without initiating instrument response.

Dead time, instrument The time that elapses while the input to an instrument varies sufficiently to pass through the dead band zone and causes the instrument to respond.

Debug To find and correct mistakes in a program.

Density Mass per unit volume such as pounds/cubic foot.

Derivative action Control action in which the rate of change of the error signal determines the amplitude of the corrective action applied. It is calibrated in time units. When subjected to a ramp change, the derivative output precedes the straight proportional action by this time.

Deviation The departure from a desired value; the system deviation that exists after the process has been stabilized.

Diaphragm The sensing element consisting of a membrane that is deformed by the pressure differential formed across it.

Dielectric constant Related to the force of attraction between two opposite charges separated by a distance in a uniform medium.

Differential pressure The difference in static pressure between two identical pressure taps.

Digital A term applied to a signal or device that uses binary digits to represent continuous values or discrete states.

Digital computer A computer operating on data in the form of digits—discrete quantities of variables rather than continuous.

Digital output An output signal that represents the size of an input in the form of a series of discrete quantities.

Digital-to-analog converter (D/A) A device or circuit that converts a digital value to an analog signal value.

Discrete A control signal that is either on or off; also referred to as a *control system.*

Disk operating system (DOS) Program used to control the transfer of information to and from a disk, such as MS DOS.

Distributed control system A system that, while being functionally integrated, consists of subsystems that may be physically separate and remotely located from one another.

Drift A change of a reading or a setpoint value over long periods due to several factors, including change in ambient temperature, time, and line voltage.

Dual element sensor A sensor assembly with two independent sensing elements.

Duplex wire A pair of wires insulated from each other and with an outer jacket of insulation around the inner insulated pair.

Duty cycle The total time to one on/off cycle. Usually refers to the on/off cycle time of a temperature controller.

Dynamic behavior Behavior as a function of time.

Electrical interference The electrical noise induced upon the signal wires that obscures the wanted information signal.

Elevation Additional or unwanted pressure on the low side of a DP cell. The term normally applies to a condition that causes the 4-mA output of a DP cell to be low when there is 0% level so the output is adjusted to zero.

Endothermic Absorbs heat. A process is said to be endothermic when it absorbs heat.

EPROM Erasable programmable read-only memory. The EPROM can be erased by ultraviolet light or electricity.

Equilibrium The condition of a system when all inputs and outputs (supply and demand) have steadied down and are in balance.

Error The difference between the actual and the true values, often expressed as a percentage of either span or upper-range value.

Excitation The external application of electrical voltage or current applied to a transducer for normal operation.

Exothermic To give off heat. A process is said to be exothermic when it gives off heat.

Fahrenheit A temperature scale defined by 32° at the ice point and 212° at the boiling point of water at sea level.

Feedback Information about the status of the controlled variable that may be compared with information that is desired in the interest of making them coincide.

Final control element Component of a control system (such as a valve) that directly regulates the flow of energy or material to the process.

Floppy disk A small, flexible disk carrying a magnetic medium in which digital data are stored for later retrieval and use.

Flow Travel of liquids or gases in response to a force (gravity or pressure).

Flow meter A device used for measuring the flow or quantity of a moving fluid.

Flow rate The actual speed or velocity of fluid movement.

FPM Flow velocity in feet per minute.

FPS Flow velocity in feet per second.

Frequency Occurrence of a periodic function (with time as the independent variable), generally specified as a certain number of cycles per unit time.

Frequency corner That frequency in the Bode diagram indicated by a breakpoint—the intersection of a straight line drawn asymptotically to the log-gain versus log-frequency curve and the unit log-gain abscissa.

Frequency response analysis A system of dynamic analysis that consists of applying sinusoidal changes to the input and recording both the input and output on the same time base using an oscillograph. By applying these data to the Bode diagram, the dynamic characteristics can be determined graphically.

Friction A facet resulting in energy loss, due to heat dissipation, when fluid is moving through a pipe. Friction results when a moving fluid comes into contact with the pipe walls. The viscosity and the fluid motion result in heat energy, which must take away the total energy of the moving fluid.

Function The purpose of, or an action performed by, a device.

Gauge pressure Absolute pressure minus local atmospheric pressure.

Gain (magnitude ratio) The ratio of change in output divided by the change in input that caused it. Both output and input must be expressed in the same units, making gain a pure (dimensionless) number.

Gain, loop The combined input/output magnitude ratios of all the individual loop components multiplied to obtain the overall gain.

Gain, margin The sinusoidal frequency at which the output/input magnitude ratio equals unity and feedback achieves a phase angle of −180°.

Gain, static (zero-frequency gain) The output/input amplitude ratio of a component or system as frequency approaches zero. Used to denote the amount of "gain" in milliamps per input change.

GPH Volumetric flow rate in gallons per hour.

GPM Volumetric flow rate in gallons per minute.

Handler A small program that handles data flow to and from specific pieces of hardware for use by the other software.

Hardcopy Output in a permanent form (usually a printout) rather than in temporary form, as on disk or display terminal.

Hardware Physical equipment, for example, mechanical, magnetic, electrical, or electronic devices. Something that you can touch with your finger.

Head loss The loss of pressure in a flow system measure using a length parameter (i.e., inches of water).

Head pressure Expression of pressure in terms of the height of fluid.

Heat Thermal energy. Heat is expressed in units of calories, or BTUs.

Heat transfer The process of thermal energy flowing from a body of high energy to a body of low energy.

Host The primary or controlling computer in a multiple-part system.

Hunting Oscillation or cycling that may be of appreciable amplitude caused by the system's persistent effort to achieve a prescribed level of control.

Hysteresis Difference between upscale and downscale results in instrument response when subjected to the same input approached from opposite directions.

Icon A graphical, functional symbol of display; a graphic representation of a function or functions to be performed by the computer.

Identification The sequence of letters or digits, or both, used to designate an individual instrument or loop.

Impedance The total opposition to current flow (resistive plus reactive).

Input Incoming signal to measuring instrument, control units, or system.

Input impedance The impedance of a meter as seen from the source. In the case of a volt meter, this impedance has to be taken into account when the source impedance is high; in the case of an ammeter, when the source impedance is low.

Instrument In process measurement and control, this term is used broadly to describe any device that performs a measuring or controlling function.

Integral control action Action in which the controller's output is proportional to the time integral of the error input. When used in combination with proportional action, it was previously called *reset action.*

Integral time The calibrated time on the controller integral (reset) dial that represents the time that will elapse while the open-loop controller repeats proportional action.

Integral windup The overcharging, in the presence of a continuous error, of the integral capacitor (bellows, in a pneumatic controller) that must discharge through a longtime, constant discharge path and that prevents a quick return to the desired control point.

Integrator Often used with a flow meter to totalize the area under the flow record; for example, gallons per minute × minutes = total gallons. It produces a digital readout of total flow.

Interface The means by which two systems or devices are connected and interact with each other.

Intrinsically safe An instrument that will not produce any spark or thermal effects under normal or abnormal conditions that will ignite a specified gas mixture.

I/P converter A device that converts an electric current (4–20 mA) to a linear pneumatic pressure (3–15 PSI).

I/O (input/output) The interface between peripheral equipment and the digital systems.

ISA International Society for Automation; formerly known as Instrument Society of America.

Joule The basic unit of thermal energy.

Junction The point in a thermocouple where the two dissimilar metals are joined.

Kelvin (symbol K) The unit of absolute of thermodynamic temperature scale based upon the Celsius scale with 100 units between the ice point and boiling point of water at sea level. (0° Celsius = 273.15° Kelvin).

Lag A delay in output change following a change in input.

Laminar flow Streamlined flow of a fluid where viscous forces are more significant than inertial forces; generally below a Reynold's number of 2,000.

LaPlace transform A transfer function that is the operational equivalent of a complex mathematical function permitting solution by simple arithmetic techniques.

Leakage rate The maximum rate at which a fluid is permitted or determined to leak through a seal. The type of fluid, the differential pressure across the seal, the direction of leakage, and the location of the seal must be specified.

Life cycle The minimum number of cycles a transducer can endure and still remain within a specified tolerance.

Limiting A boundary imposed on the upper or lower range of a variable (for example, the pressure in a steam boiler as limited by a safety valve).

Line pressure Pressure measured at any point of a moving fluid usually referenced to atmospheric pressure.

Linearity The closeness of a calibration curve to a specified, straight line. Linearity is expressed as the maximum deviation of any calibration point on a specified, straight line during any one calibration cycle.

Load impedance The impedance presented to the output terminals of a transducer by the associated external circuitry.

Local The location of an instrument that is neither in nor on a panel or console, nor is it mounted in a control room. The term "field" is often used for "local."

Local panel A panel that is not a central or main panel.

Log gain Gain expressed on a logarithmic scale.

Loop A signal path; a combination of two or more instruments or control functions arranged so that signals pass from one to another for the purpose of measurement and/or control.

Loop resistance The total resistance of a thermocouple circuit caused by the resistance of the thermocouple wire.

Manipulated variable That which is altered by the automatic control equipment so as to change the variable under control and make it conform with the desired value.

Manual loading station A device or function having a manually adjusted output that is used to actuate one or more remote devices.

Manual reset (adjustment) The adjustment on a proportioning controller that shifts the proportioning band in relationship to the setpoint to eliminate droop or offset errors.

Mass flow rate Volumetric flow rate times density (i.e., pounds per hour).

Mass storage A device like a disk or magnetic tape (magtape) that can store large amounts of data readily accessible to the central processing unit.

Maximum operating temperature The maximum temperature at which an instrument or sensor can be operated safely.

Maximum power rating The maximum power in watts that a device can handle safely.

Mean The average of the maximum and the minimum of a process equilibrium.

Measurement The determination of the existence or the magnitude of a variable.

Measuring element An element that converts any system activity or condition into a form or language that the controller can understand.

Memory Pertaining to that storage device in which programs and data are stored and easily obtained by the CPU for execution.

Meter run A field term used to describe the upstream and downstream piping, including orifice and orifice flanging. The straight runs of piping upstream and downstream allow the flow to stabilize prior to passing through a flow orifice while the down-stream run allows the flow to exit the orifice with a minimum of back pressure.

Microamp 1/1,000,000 of an ampere (mA).

Milliamp 1/1,000 of an ampere (mA).

Mineral-insulated thermocouple A type of thermocouple cable that has an outer metal sheath and mineral (magnesium oxide) insulation inside separating a pair of thermocouple wires from themselves and the outer sheath. It is ideally suited for high-temperature and severe duty applications.

Monitor A general term for an instrument or instrument systems used to measure or sense the status or magnitude of one or more variables for the purpose of deriving useful information.

NEMA-4 A standard from the National Electrical Manufacturers' Association; it defines enclosures intended for indoor and outdoor use primarily to provide a degree of protection against windblown dust and rain, splashing water, and hose-directed water.

NEMA-7 A standard from the National Electrical Manufacturers' Association; it defines explosion-proof enclosures for use in locations classified as Class I, groups A, B, C, or D, as specified by the *National Electrical Code®*.

NEMA-12 A standard from the National Electrical Manufacturers' Association; it defines enclosures with protection against dirt, splashes with noncorrosive liquids, and salt spray.

Nichols diagram (Nichols chart) A plot of magnitude and phase contours of return-transfer function referred to ordinates of logarithmic loop gain and abscissae of loop-phase angel.

Noise Unwanted signal components that obscure the genuine signal information that is being sought (electrical interference).

NPT National pipe thread.

OD Outside diameter.

Off-line (1) Pertaining to equipment or programs not under control of the computer; (2) pertaining to a computer that is not actively monitoring or controlling a process.

Offset The difference between what we get and what we want—the difference between the point at which the process stabilizes and the instruction introduced into the controller by the setpoint.

On-line (1) Pertaining to equipment or programs under control of the computer; (2) pertaining to a computer that is actively monitoring or controlling a process or operation.

Open loop Control without feedback, for example, an automatic washer.

Operating system   A collection of programs that controls the overall operation of the computer and performs such tasks as assigning places in memory to programs and data, processing interrupts, scheduling jobs, and controlling the overall input/output of the system.

Optimum   The highest obtainable proficiency of control; for example, supply equals demand, and offset has been reduced to a minimum (hopefully zero).

Orifice flange   A circular, bolted pipe flange designed to accept an orifice plate for flow measurement. Orifice flanges are usually thicker than regular flanges to allow for taping (for differential pressure measurements).

Orifice plate   A circular plate with an outside diameter slightly smaller than the associated orifice flange to allow clearance for flange bolts. The plate has a hole of set diameter in the middle so that when a fluid moves through it, a differential pressure results that can be measured. Orifice plates are also equipped with a TAG as well as stamped to show the direction of fluid flow.

Oscillograph recorder   A device that makes a high-speed record or electrical variations with respect to time; for example, an ordinary recorder might have a chart speed of 3/4 inch per hour while an oscillograph could have a chart speed of 3/4 inch per second or faster.

Output   The signal provided by an instrument; for example, the signal that the controller delivers to the valve operator is the controller output.

Overdamped   Damped so that overshoot cannot occur.

Overshoot   The persistent effort of the control system to reach the desired level, which frequently results in going beyond (overshooting) the mark.

Panel   A structure that has a group of instruments mounted on it, houses an operator-process inerface, and is chosen to have an unique designation.

Panel-mounted   A term applied to an instrument that is mounted on a panel or console and is accessible for an operator's normal use.

Parallax   An optical illusion that occurs in analog meters and causes reading errors. It occurs when the viewing eye is not in the same plane, perpendicular to the meter face, as the indicating needle.

Peripheral   A device that is external to the CPU and main memory (printer), but is connected by the appropriate electrical connections.

Phase   The condition of a periodic function with respect to a reference time.

Phase difference   The time, usually expressed in electrical degrees, by which one wave leads or lags another.

P&ID   Process and instrumentation diagram.

Pilot light   A light that indicates in which of a number of normal conditions of a system or device exists.

Pixel   Picture element. Definable locations on a display screen that are used to form images on the screen. For graphic displays, screens with more pixels provide higher resolution.

Port   A signal input (access) or output point on a computer.

Potentiometer   (1) A variable resistor often used to control a circuit. (2) A balancing bridge used to measure voltage.

Power supply   A separate unit or part of a circuit that supplies power to the rest of the circuit or system.

PPM   Parts per million; sometimes used to express temperature coefficients.

Precision   The ability of an instrument to provide the same output signal given the identical input.

Primary element   Synonym of *sensor*.

Probe   A generic term that is used to describe many types of temperature sensors.

Process   The variable for which supply and demand must be balanced—the system under control, excluding the equipment that does the controlling.

Process meter   A panel meter with sizable zero and span adjustment capabilities that can be scaled for readout in engineering units for signals such as 4–20 mA, 10–50 mA, 1–5 v DC.

Process variable   Any variable property of a process; the part of the process that changes and therefore needs to be controlled.

Program   A series of instructions that logically solve given problems and manipulate data.

Programmable logic controller   A controller, usually with multiple inputs and outputs, that contains an alterable program.

PROM (Programmable read-only memory)   A semiconductor whose contents cannot be changed by the computer after it has been programmed.

Proof pressure   The specified pressure that may be applied to the sensing element of a transducer without causing a permanent change in the output characterics.

Proportional band   The reciprocal of gain expressed as a percentage. Refers to the percentage of the controller's span of measurement over which the full travel of the control valve occurs.

Proportional control   Control action in which there is a fixed gain or attenuation between output and input.

Proportional, integral, derivative (PID)   A three-mode control action in which the controller has time-proportioning, integral (auto reset), and derivative rate actions.

Protection head   An enclosure usually made out of metal at the end of a heater or probe where connections are made.

Protection tube A metal or ceramic tube, closed at one end, into which a temperature sensor is inserted. The tube protects the sensor from the medium into which it is inserted.

Protocol A formal definition that describes how data are to be exchanged.

PSIA Pounds per square inch absolute. Pressure referenced to a vacuum.

PSID Pounds per square inch differential. Pressure difference between two points.

PSIG Pounds per square inch gage. Pressure referenced to a standard atmosphere.

Ramp An increase or decrease of the variable at a constant rate of change.

Random access memory (RAM) Memory that can be read and changed during computer operation. Unlike other semiconductor memory, RAM is volatile. If power to the RAM is disrupted or lost, all stored data are lost.

Range Those values over which a transmitter is intended to measure.

Rankine (°R) An absolute temperature scale based on the Farenheight scale with 180° between the ice point and the boiling point of water. (459.67°R = 0°F).

Rate action That portion of controller output that is proportional to the rate of change of input; see *Derivative action.*

Reaction curve In process control, a reaction curve is obtained by applying a step change (either in load or setpoint) and plotting the response of the controlled variable with respect to time.

Read-only memory (ROM) Memory that contains fixed data. The computer can read the data but cannot change the data in any way.

Real-time clock A device that automatically maintains time in conventional time units for use in program execution and event initiation.

Recovery time The length of time it takes a transducer to return to normal after a proof-pressure has been applied.

Reference junction The cold junction in a thermocouple circuit that is held at a stable known temperature. Usually 0°C (32°F), but any reference temperature can be used.

Relay A device whose function is to pass on information in an unchanged form or in some modified form.

Repeatability The ability of a transmitter to reproduce output readings when the same measured value is applied to it consecutively, under the same conditions, and in the same direction. Repeatability is expressed as the maximum difference between two readings.

Reproducibility The exactness with which a measurement or other condition can be duplicated over time.

Reset action See *integral control action.*

Reset time See *integral time.*

Reset windup See *integral windup.*

Resistance An opposition to flow that accounts for the dissipation of energy and limits flow. Flow from a water tap, for example, is limited to what the available pressure can push through the tap opening electrical resistance.

$$(\text{ohms}) = \frac{\text{potential (expressed in volts)}}{\text{flow (expressed in amperes)}}$$

Resolution The smallest detectable increment of measurement. Resolution is usually defined by the smallest number of bits used to define a reading (measurement).

Resonant frequency (instrument) The measured frequency in which a transducer responds with maximum amplitude.

Response Reaction to a forcing function applied to the input. The variation in measured variables that occurs as the result of step sinusoidal, RAM, or other kind of input.

Response time (time constant) The time required by a sensor to reach 63.2% of a step change under a specified set of conditions. Five time constants are required for the sensor to stabilize at 100% of the step change value.

Reynold's number The ratio of inertial and viscous forces in a fluid defined by the formula:

$$Re = p \times \frac{VD}{\mu}$$

where p = density of fluid, μ (mu) = viscosity in centipoise (CP), V = velocity, and D = inside diameter of pipe.

(Note: This is the basic Reynold's number equation. This equation will yield the same results as the equation presented in Chapter 5 of the text.)

Routine A small program used by many other programs to perform a specific task.

RTD Resistance temperature detector.

Scan To sample, in a predetermined manner, each of a number of variables intermittently.

Secondary device A part of a flow meter that receives a signal proportional to the flow rate, from the primary device and displays, records, and/or transmits the signal.

Self-heating Internal heating of a transducer as a result of power dissipation.

Self-regulation The ability of an open-loop process or other device to settle out (stabilize) at some new operating plant after a load change has taken place.

Sensing element The part of a transducer that reacts directly in response to the input.

Sensitivity The minimum change in an input signal to which an instrument can respond.

Sensor That part of a loop or instrument that first senses the value of a process variable. The sensor is also known as a *detector* or *primary element.*

Servo techniques The mathematical and graphical methods devised to analyze and optimize the behavior of control systems.

Setpoint The desired value where the process should be maintained.

Shared controller A controller containing preprogrammed algorithms that are usually accessible, configurable, and assignable. Permits a number of process variables to be controlled by a single device.

Shared display The operator interface device (usually a video screen) used to display process control information from a number of sources at the command of the operator.

Signal Information in the form of a pneumatic pressure, and electric current or mechanical position that carries information from one control loop component to another.

Software The collection of programs and routines associated with a computer.

Span The difference between the upper and lower limits of a range expressed in the same units as the range.

Span adjustment The ability to adjust the gain of a process or instrument so that a specified display span in engineering units corresponds to a specified signal span. For instance, a display of 200°F may correspond to the 16-mA span of a 4–20 mA transmitter signal.

Span error Errors identified as an output signal that does not reflect 100% of the output signal or does not follow the input span.

Specific gravity The ratio of mass of any material to the mass of the same volume of water at 4°C.

Specific heat The ratio of thermal energy required to raise the temperature of a body 1° to the thermal energy required to raise an equal mass of water 1°.

Stability That desirable condition in which input and output are in balance and will remain so unless subjected to an external stimulus.

Static behavior Behavior that is either not a function of time or that takes place over a sufficient length of time that dynamic changes become of minor importance.

Static pressure Pressure of a fluid whether in motion or at rest. It can be sensed in a small hole drilled perpendicular to and flush with the flow boundaries so as not to disturb the flow boundaries in any way.

Steady flow A flow rate in the measuring section of a flow line that does not vary significantly with time.

Steady state A state in which static conditions prevail and all dynamic changes may be assumed completed.

Step change A change from one level to another in supposedly zero time.

Strain gage A measuring element for converting force, pressure, tension, etc. into an electrical signal.

Summing point A point at which several signals can be added algebraically.

Suppression Lowering the output caused by an additional or unwanted pressure on the high side of a DP cell. The term is normally used for a condition that has the 4-mA output high when there is 0% level.

Surge current A current of a short duration that occurs when capacitive power is first applied to capacitive loads or temperature-dependent resistive loads. Usually not lasting more that several cycles.

Switch A device that connects, disconnects, selects, or transfers one or more circuits and is not designated as a controller, a relay, or control valve.

Syntax The rules governing the structure of a language.

System Generally refers to all control components, including process, measurements, controller, operator, and valves, along with any other additional equipment that may contribute to its operation.

Tag Number Alphanumeric sequence that identifies a device by a unique identifier.

Temperature range, operable The range of ambient temperatures, given by their extremes, within which a transducer may be operated. Exceeding compensated range may require recalibration.

Terminal A device for operator-machine interface; for example, CRTs, typewriters, teletypes with keyboard input, or telephone modems.

Test point A process connection in which no instrument is permanently connected, but is intended for the temporary or intermittent connection of an instrument.

Thermal conductivity The property of a material to conduct heat in the form of thermal energy.

Thermal expansion An increase in size due to an increase in temperature expressed in units of an increase in length or increase in size per degree.

Thermocouple A device constructed of two dissimilar metals that generates a small voltage as a function of temperature difference between measuring and reference junctions. The voltage can be measured and its magnitude used as a measure of the temperature in question.

Thermowell   A closed-end tube designed to protect temperature sensors (thermocouples) from harsh environments.

Time constant   The product of resistance × capacitance ($t = RC$), which becomes the time required for a first-order system to reach 63.2% of a total change when forced by a step. In so-called "high-order" systems, there is a time constant for each of the first-order components.

Transducer   A device that converts information of one physical form to another physical type in its output (e.g., a thermocouple converts temperature into millivoltage).

Transfer function   A mathematical description of the output divided by input relationship that a component or a complete system exhibits. It often refers to the LaPlace transform of output over the LaPlace transform of input with zero initial conditions.

Transmitter   A device that senses a process variable through the medium of a sensor and has an output whose steady-state value varies only as a predetermined function of the process variable. The sensor may or may not be integral with the transmitter.

Transmitter (2-wire)   A device that is used to transmit temperature data from either a thermocouple or an RTD via a two-wire current loop. The loop has an external power supply and the transmitter acts as a variable resistor with respect to its input signal.

Transportation lag   A delay caused by the time required for material to travel from one point to another; for example, water flowing in a pipe at 10 feet per second requires 10.0 seconds to travel 100 feet, and if this 100 feet exists between manipulation and measurement, it would constitute at 10-second lag.

Turbulent flow   When forces due to inertia are more significant than forces due to viscosity. This typically occurs with a Reynold's number in excess of 4,000.

Ultimate period   The time period of one cycle at the natural frequency of the system where it is allowed to oscillate without damping.

Vacuum pressure   Any pressure less than atmospheric pressure.

Value   The level of the signal being measured or controlled.

Variable   A level, quantity, or other condition that is subject to change. This may be regulated (the controlled variable) or simply measured (a barometer measuring atmospheric pressure).

Velocity   The time rate of change of displacement $\left(\dfrac{\Delta x}{\Delta t}\right)$

Vena contracta   A term used to describe the point downstream of an orifice plate where the fluid velocity is greatest and pressure is lowest due to the inertia of the moving fluid.

Viscosity   The inherent resistance of a substance to flow.

Volume flow rate   Calculated using the area of the full closed conduit and the average fluid velocity in the form, $Q = VA$, to arrive at the total volume quantity of flow. Q = volumetric flow rate, V = average fluid velocity, and A = cross-sectional area of the pipe.

Word   Number of bits treated as a single unit by the CPU. In an 8-bit machine, the word length is 8-bits; in a 16-bit machine, the word length is 16-bits.

Zero adjustment   The ability to adjust the display of a process or meter so that a zero on the display corresponds to a nonzero signal, such as 4 mA. The adjustment is normally expressed in counts.

Zero frequency gain   Static gain or change in output divided by the change in input that caused it, after sufficient time has elapsed to eliminate the dynamic behavior components.

Zero offset   The difference expressed in degrees between true zero and an indication given by a measuring instrument.

Zero shift   Change resulting from an error that is the same throughout the scale.

Zero suppression   The span of an indicator or chart recorder may be offset from zero (zero suppressed) such that neither limit of the span will be zero.

Index